T0393164

Artificial Intelligence in Higher Education

The global adoption of technology in education is transforming the way we teach and learn. Artificial Intelligence is one of the disruptive techniques to customise the experience of different learning groups, teachers, and tutors. This book offers knowledge in intelligent teaching/learning systems, and advances in e-learning and assessment systems.

Features include:

- Highlights the broad field of AI applications in education, regarding any type of Artificial Intelligence that is correlated with education.
- Discusses learning methodologies, intelligent tutoring systems, intelligent student guidance and assessments, intelligent educational chatbots, and artificial tutors.
- Presents the practicality and applicability implications of AI in education.
- Includes new and current research from research centers and higher education universities.
- Offers case studies of AI techniques in educational activities.

Artificial Intelligence in Higher Education: A Practical Approach will find interest with academicians which includes teachers, students of various disciplines, higher education policymakers who believe in transforming the education industry, research scholars who are pursuing their PhD or Post Doctorate in the field of Education Technology, Education, and Learning, and so on, and those working in the areas of Education Technology and Artificial Intelligence such as industry professionals in education management and e-learning companies.

Artificial Intelligence in Higher Education

A Practical Approach

Edited by

**Prathamesh Churi, Shubham Joshi,
Mohamed Elhoseny, and Amina Omrane**

CRC Press
Taylor & Francis Group
Boca Raton London

CRC Press is an imprint of the
Taylor & Francis Group, an **informa** business

First edition published 2023
by CRC Press
6000 Broken Sound Parkway NW, Suite 300, Boca Raton, FL 33487–2742

and by CRC Press
4 Park Square, Milton Park, Abingdon, Oxon, OX14 4RN

CRC Press is an imprint of Taylor & Francis Group, LLC

© 2023 Taylor & Francis Group, LLC

Library of Congress Cataloging-in-Publication Data

Names: Churi, Prathamesh, editor. | Joshi, Shubham, editor. | Elhoseny, Mohamed, editor. | Omrane, Amina, editor.
Title: Artificial Intelligence in higher education : a practical approach / edited by Prathamesh Churi, Shubham Joshi, Mohamed Elhoseny, and Amina Omrane.
Description: First Edition. | Boca Raton, FL : CRC Press, 2022. | Includes bibliographical references and index.
Identifiers: LCCN 2021059239 | ISBN 9781032026060 (Hardcover) | ISBN 9781032026077 (Paperback) | ISBN 9781003184157 (eBook)
Subjects: LCSH: Artificial Intelligence—Educational applications. | Education, Higher—Computer-assisted instruction. | Intelligent tutoring systems.
Classification: LCC LB1028.43 .A7897 2022 | DDC 378.1/7344678—dc23/eng/20220209
LC record available at https://lccn.loc.gov/2021059239

ISBN: 978-1-032-02606-0 (hbk)
ISBN: 978-1-032-02607-7 (pbk)
ISBN: 978-1-003-18415-7 (ebk)

DOI: 10.1201/9781003184157

Typeset in Times
by Apex CoVantage, LLC

This book is dedicated to all the Teachers, Students of Higher Education and Researchers in Education Technology and ARTIFICIAL INTELLIGENCE field

Contents

Preface

We are pleased to present to you our first edition of this book, *Artificial Intelligence in Higher Education: A Practical Approach*. The application of Artificial Intelligence has rapidly increased in many sectors including healthcare, e-commerce, banking, and supply chain management. The development of Artificial Intelligence-based educational techniques has been upgraded significantly over the past few years. Implementing Artificial Intelligence and artificial neural networks in education includes many kinds of intelligent instructional and evaluation techniques such as intelligent tutoring systems, intelligent assessment of student performance, intelligent virtual agents, talking robots, humanized chatbots, and any other educational technique based on Artificial Intelligence. Global adoption of technology in education is transforming the way we teach and learn. Artificial Intelligence is one of the disruptive techniques to customize the experience of different learning groups, teachers, and tutors. This book will provide knowledge regarding intelligent teaching-learning systems as well as advances in e-learning systems and assessment systems.

In the 1970s AIED (Artificial Intelligence Education) emerged as a specialist area covering new technology in teaching and learning, mainly in higher education. AIED aimed to provide more personalized, flexible, inclusive, and engaging learning, also automate daily learning tasks through automated assessment and feedback. In theory, AIED could help parents improve early linguistic development for their children, also help teachers choose tools, organize classes, increase participation, and personalize teaching for their students. AIED is contained as a robot or virtual assistance (Vas), and it combined virtual reality. It served as sensors that captured visual, auditory, and physiological data of students and teachers. This data type of learning can further understanding of how learning occurs in real time, and help teachers choose powerful approaches to teaching. AIED was purported to be capable of developing tools that help combat student or teacher burnout, and that may help eliminate the gap in achievement between students due to individual or social differences. However, despite decades of research, the tools developed thus far by AIED have not made full use of potential technology and seem far from fulfilling these promises.

This book is intended to enlighten the thrust of novel technological interventions in education and learning outcomes to foster pedagogies through Artificial Intelligence-based techniques to improve education needs that schematize better perspectives at large. The primary aim of this edited book is to help researchers, academicians, and educators develop principles for the design of computer-based learning systems. Its premise is that such principles involve the modeling and representation of relevant aspects of knowledge, before implementation or during execution, and hence require the application of AI techniques and concepts.

This book aims to highlight the broad field of Artificial Intelligence applications in education, regarding any type of Artificial Intelligence that is correlated with education, such as learning methodologies, intelligent tutoring systems, intelligent student guidance and assessments, intelligent educational chatbots, artificial tutors, and so on, in order to advance and enrich the existing literature with new Artificial Intelligence approaches and methodologies in education.

The book contains 12 chapters written by 32 contributors from nine different countries which include India, Pakistan, Malaysia, Saudi Arabia, Russia, Spain, Swaziland, South Africa, and Malta. We are sure that this book will encourage researchers in this field.

Prof. Prathamesh Churi
Assistant Professor, Mukesh Patel School
of Technology Management and Engineering,
NMIMS University, Mumbai, India

Shubham Joshi
Assistant Professor, Department of
Computer Engineering, Mukesh Patel School of Technology
Management & Engineering, Shirpur, India

Dr. Mohamed Elhoseny
Assistant Associate Professor,
University of Sharjah, United Arab Emirates

Dr. Amina Omrane
Associate Professor of management (authorized to supervise researches
in management science and entrepreneurship) at FSEG Faculty
(Faculty of economic sciences and management),
University of Sfax, Tunisia

Acknowledgements

First and foremost, praises and thanks to God, the Almighty, for His showers of blessings throughout our work to successfully complete the book *Artificial Intelligence in Higher Education: A Practical Approach.*

The contributing authors have been a real motivation and key in establishing CRC Press/Taylor & Francis Group as one of the best publishers in the subject of Data Protection and Privacy. We thank them all for considering and trusting our book for publishing their valuable work. We also thank all authors for their kind co-operation extended during the various stages of production of the book with CRC Press/Taylor & Francis.

For the success of any edited book, reviewers are an essential part and hence reviewers merit sincere appreciation. Reviewers' input are used to improve the quality of submitted book chapters. The reviewing of a book chapter is essential to assure the quality of the chapter published in any book. We thank the following reviewers for their excellent contributions during the review process:

- Dr. Prashant Gupta, Maharashtra Institute of Technology, Aurangabad, India
- Dr. Antonio José Moreno Guerrero, University of Granada, Spain
- Prof. Ameyaa Biwalkar, MPSTME, NMIMS University, India
- Dr. Sheshadri Chatterjee, Postdoctoral Researcher, Department of Computer Science and Engineering, IIT Kharagpur
- Dr. Astha Bhanot, Business Administration Department, College of Business & Administration, Princess Nourah bint Abdulrahman University, Riyadh, Kingdom of Saudi Arabia
- Dr. Upasana Gitanjali Singh, University of KwaZulu-Natal, South Africa

The overwhelming response from the authors across the world has been a real motivation and support in moving this book forward in the area of AI in higher education. Last but not least, we would like to thank Ms. Cindy Renee Carelli (Executive Editor—Engineering, CRC Press/Taylor & Francis) and Ms. Erin Harris (Senior Editorial Assistant, Engineering, CRC Press/Taylor & Francis) for their valuable support in the editing process.

Editors

Prof. Prathamesh Churi, SMIEEE, MACM, LMCSI, LMISTE, MIGIP is a faculty member in the Computer Engineering Department in the School of Technology Management and Engineering at NMIMS University, India. He is a senior member of IEEE. He is currently serving as an Associate Editor of *International Journal of Advances in Intelligent Informatics* (Scopus), *International Journal of Innovative Teaching and Learning in Higher Education* and *International Journal of Information Security and Privacy* (Scopus, ESCI, ABDC-C). He is also a research mentor for the company Cerebranium in Germany. He is actively involved in peer review process of reputed IEEE and Springer journals such as *IEEE Transactions on Education, Springer Education and Information Technologies* and 17 other journals. He has published more than 60 research papers in national/international conferences and journals (Scopus, ESCI and SCI Indexed). Prof. Churi has five patents (including two Australian patents) in the fields of wireless sensor networks, machine learning and Internet of Things, and privacy preserving key management protocol. He has edited three international books (CRC Press/Taylor & Francis publications) in the fields of data privacy and education technology. He has been a keynote speaker, chair, and convener in international conferences including such flagship conferences as IEEE TALE 2017–2020 and Springer ICACDS, among others. He recently received Best Young Researcher Award from GISR Foundation for his research contribution to the field of data privacy and security, education technology. He has also earned an appreciation award for best faculty at NMIMS University. He is the recipient of the Excellent Reviewer Award from *Springer Journal of Educational Technology Research and Development* in 2021. Prof. Churi is an active leader, coach, mentor, and volunteer in many non-profit organizations. He is also involved as a board of study member in many universities for curriculum development and educational transformations. His relaxation and self-development lay in pursuing his hobbies which mainly include expressing views be it in public writing columns or blogging.

Linked in profile: www.linkedin.com/in/prathameshchuri/

Dr. Shubham Joshi is a valued IEEE senior member. He has 14 years of experience in teaching, research and consulting. He is a visionary researcher passionate about a new era of creation. With the advent of a novel platform for learning and development, Dr. Joshi has served for many aspirants, technology savvy people, working professionals, and organizations. He has authored more than 40 research papers, with 43 citations (Google Scholar), an h index of 3; i-index-1, and has 5 Scopus publications (2012 onwards). He has presented research papers and delivered talks in more than 10 conferences since 2010. He presented his research paper at the IEEE International Conference in Bali, Indonesia. He has mentored NBA and NAAC accreditation processes since 2012. In 2009 he authored a book on web technology and programming (Shiva Publication, Indore). He has one patent granted on Blood Sugar Index (1 January 2016, Indian Patent Organization). He received Outstanding Researcher Award by Sbyte Technologies on 25 July 2020, earned professional membership of the British Computer Society for 2015, became ISACA Academic Advocate Member from August 2015 onwards, and has been a member of the Computer Society of India since April 2016. Dr. Joshi founded two start-up ventures in 2014 and 2016, namely Perception & Endeavours Pune and Quodra Noida.

Dr. Mohamed Elhoseny, PhD, is an Associate Professor at Mansoura University, Egypt. Dr. Elhoseny is an ACM Distinguished Speaker and IEEE senior member. His research interests include smart cities, network security, Artificial Intelligence, the Internet of Things, and intelligent systems. Dr. Elhoseny is the founder and Editor-in-Chief of *IJSSTA* journal published by IGI Global. Also, he is an Associate Editor at several Q1 journals. Moreover, he served as the co-chair, the publication chair, the programme chair, and a track chair for several international conferences held by recognized publishers such as IEEE and Springer. Dr. Elhoseny is the Editor-in-Chief of the Studies in Distributed Intelligence ASPG book series, the Editor-in-Chief of The Sensors Communication for Urban Intelligence CRC Press/ Taylor & Francis book series, and the Editor-in-Chief of The Distributed Sensing and Intelligent Systems CRC Press/Taylor & Francis book series. He was granted several awards by diverse funding bodies such as the Egypt State Encouragement Award in 2018, the Young Researcher Award in Artificial Intelligence from the Federation of Arab Scientific Research Councils in 2019, Obada International Prize for Young Distinguished Scientists 2020, Mansoura University Young Researcher Award 2019, the SRGE Best Young Researcher Award in 2017, and the best PhD thesis in Mansoura University in 2015.

Dr. Amina Omrane is an Associate Professor authorized to supervise researchers in management science at the University of Sfax, as well as a researcher at ECSTRA research center, based in IHEC-Carthage, Tunisia. She has over 12 years of academic experience in management science and entrepreneurship, and teaches postgraduate and masters' courses related to research methods, strategic management, business plans, and entrepreneurship, as well as human/personal development. She earned her master's degree in management by the goals from ISGI-Sfax, as well as one in Management and Strategy from IHEC-Carthage, Tunisia, before earning her PhD in management science from the University of Jean Moulin (Lyon III) and IHEC-Carthage. Dr. Omrane has presented numerous research papers at international conferences/congresses and has published many other papers in international journals such as *IJBG, IJESM, IJBE, JAB, TCR, La RIPME, IJMP, FIIB Business Review*, and *La Revue des Sciences de Gestion*. She has served as a reviewer for many other international journals. She has authored five books on management science, entrepreneurship, and human development, as well as four book chapters, and has also coedited two collective books on sustainable entrepreneurship, renewable energy, and digitalization, higher education, and new business models. Dr. Omrane is a guest editor for reputed Scopus and ABDC journals, and has also served as chair, moderator, and speaker for many conferences, workshops, and event sessions. Many of her research and scientific projects are currently under consideration and revolve around SME development, innovation, and sustainability issues.

As a researcher at ECSTRA research center in IHEC-Carthage, she has also coedited and published numerous books, as well as many papers in peer-reviewed and ranked journals and international conference proceedings. Finally, Dr. Omrane is also serving as a reviewer for many other Emerald, T&F, and Inderscience journals.

Contributors

Donnie Adams
University of Malaya
Malaysia

Alla L. Arkhangelskaya
Russian Language Institute
Peoples' Friendship University
 of Russia
Moscow, Russia

Muhammad Mujtaba Asad
Sukkur IBA University
Pakistan

Roha Athar
Sukkur IBA University
Sukkur, Pakistan

Archana Bhise
MPSTME, NMIMS University
India

Patrick Camilleri
Faculty of Education
University of Malta
Malta

Sheshadri Chatterjee
Department of Computer Science and
 Engineering
IIT Kharagpu
India

Ranjan Chaudhuri
Marketing Department
NITIE
Mumbai, India

Kee-Man Chuah
University of Malaysia
Sarawak, Malaysia

Al-Karim Datoo
Sukkur IBA University
Sukkur, Pakistan

Suvojit Dhara
Department of Mathematics
IIT Kharagpur
India

Pablo Dúo-Terrón
CEIP Príncipe Felipe
Spain

Soumya Kanti Ghosh
Department of Computer Science and
 Engineering
IIT Kharagpur
India

Adrijit Goswami
Department of Mathematics
IIT Kharagpur
India

Prashant Gupta
Maharashtra Institute of Technology
Aurangabad, India

Salma Idrees
Sukkur IBA University
Pakistan

Imran Khan
Department of English
College of Arts
University of Ha'il
Ha'il, Saudi Arabia

Trishul Kulkarni
Maharashtra Institute of Technology
Aurangabad, India

Jesús López-Belmonte
University of Granada
Spain

Natalia S. Makarova
Peoples' Friendship University of
  Russia
Moscow, Russia

José-Antonio Marín-Marín
University of Granada
Spain

Ramsha Mazhar
Allan Gray Proprietary Ltd
Cape Town, South Africa

Antonio-José Moreno-Guerrero
University of Granada
Spain

Ami Munshi
MPSTME, NMIMS University
India

Irfan Ahmed Rind
Sukkur IBA University
Sukkur, Pakistan

Anjana Rodrigues
MPSTME, NMIMS University
India

Olga I. Rudenko-Morgun
Russian Language Institute
Peoples' Friendship University of Russia
Moscow, Russia

Zafarullah Sahito
Sukkur IBA University
Sukkur, Pakistan

Vidya Sawant
MPSTME, NMIMS University
India

Fahad Sherwani
National University of Computer and
  Emerging Sciences
Karachi, Sindh, Pakistan

Bhagwan Toksha
Maharashtra Institute of Technology
Aurangabad, India

Hena Yasmin
Department of Science
Sifundzani, Swaziland

1 AI in Education
A Few Decades from Now

Hena Yasmin, Ramsha Mazhar

CONTENTS

DOI: 10.1201/9781003184157-1

1.0 PURPOSE OF THE CHAPTER

The AI challenge is not just about educating more AI and computer experts, although that is important. It is also about building skills that AI cannot emulate. These are essential human skills such as teamwork, leadership, listening, staying positive, and dealing with people and managing crises and conflict.

Financial Times, 2017

This chapter aims to identify which knowledge and skills will remain for humans in an era of increasing Artificial Intelligence. Some of the issues surrounding the use of AI in education have been addressed, and how AI can be harnessed to improve the education and opportunities of students as they prepare to enter the modern, post-COVID-19 workforce. The need for students, employees, and society to develop the awareness and understanding that they will need to be effective, engaged, and active citizens in a world in which AI will play an increasing role is also stressed. Keeping all this in mind, we therefore need to proceed diligently and prudently towards a new educational environment where AI is used to support learners and teachers, and where we also prepare learners for a future world in which AI plays an increasing role.

1.1 WHAT IS ARTIFICIAL INTELLIGENCE (AI)

The global adoption of technology in education is transforming the way we teach and learn. Artificial Intelligence (AI) is one of the disruptive techniques to customize the experience of different learning groups, teachers, and tutors (Lisa Plitnichenko, 2020). On the other hand, it has brought three dimensional wonderful transitions in the education sector, benefiting students, teachers, and institutions. Switching to digital is a unique opportunity for the development and growth of creative youngsters, who love exploration. Digital technologies play a role in attaining success in the education of our future citizens by offering them innovative ways to learn, generate, communicate, share, and collaborate. Everyone wants change; our learners deserve change too. Artificial Intelligence is now the new normal. We are surrounded by this technology from automatic parking systems, smart sensors for taking spectacular photos, and personal assistance. Similarly, Artificial Intelligence in education has been underway for years, and is now changing traditional methods drastically as skepticism around it reduces.

The way people learn is changing all over the globe, as educational materials are becoming accessible to all through smart devices and computers. Today, students don't need to attend physical classes to study as long as they have smartphones, tablets, computers and internet connections. Let us see; what is AI?

Artificial Intelligence is a branch of computer science that creates an "artificial intelligent brain" in the form of machines to work and react something like the human brain (Karsenti, 2019). We can say it performs the functions of an intelligent or gifted child. These machines can do everything from processing data, making patterns and models, to identifying problems, reasoning, planning, solving identified problems, making predictions, and manipulating objects. The main objective of AI is to improve monotonous procedures, more speedily and more efficiently. Looking into this, AI tools can be put under the umbrella of three basic principles (Buckingham, 2018):

- **Learning:** Acquiring and processing the new experience, creating new behaviour models
- **Self-correction:** Refining the algorithms to ensure the most accurate results
- **Reasoning:** Picking up the specific algorithms to resolve a specific task

These tasks are not new; the only difference, they are performed much faster due to AI. Smartphones, tablets, and web browsers are used frequently by teachers and students to track fast pieces of information. Recognition of speech, translating languages, performing commands, tracking a person's behaviour, offering information and products that may interest consumers, providing reasoning, planning, and learning, are not a nightmare anymore. Moving further, AI is popular in gaming (for example, a computer performs an opponent's role while playing chess), car assembly (self-driving automobiles), and many other industries. Thus, there is a huge discrepancy between what AI means and its role. There have been dramatic scientific innovations, especially in machine learning, and AI incorporates machine learning. This mode of learning allows learners to unlock their abilities and skills such as intellect, acuity, problem-solving, planning, time management, and more. Now looking at this advanced AI system, it is fair to inquire how education will anticipate upcoming changes and what extra benefits teachers and learners can anticipate. Let's briefly look into the definition given by different intellects.

1.1.1   DEFINITION OF AI

AI is developing rapidly, and there are no universally accepted definitions for terms that will be important to this report (Parnas, 2017). One of the first and still most influential definitions of AI was established by McCarthy: "The study of AI is to proceed based on the conjecture that every aspect of learning or any other feature of intelligence can in principle be so precisely described that a machine can be made to simulate it" (Popenici & Kerr, 2017; Russell & Norvig, 2010).

Contemporary definitions of AI differ in various aspects and the problems to formulate a united definition of AI are caused by both constant shifts in what AI includes (Luckin et al., 2016) as well as by the interdisciplinarity of its research (AI has been studied not only by computer science, but also by philosophy, anthropology, biology, pedagogy, psychology, linguistics, cognitive science, neuroscience, statistics, and others).

One group of definitions sees AI as machines, computers or computer systems that imitate cognitive functions that are normally associated with the human mind, such as learning and problem solving (Russell & Norvig, 2010). Another group of definitions considers AI as a specific set of skills of computers; Baker and Smith (2019) define AI as "computers which perform cognitive tasks, usually associated with human minds, particularly learning and problem-solving".

The Encyclopaedia Britannica states that AI is "the ability of a digital computer or computer-controlled robot to perform tasks commonly associated with intelligent beings", where intelligent beings are those that can adapt to changing circumstances. Other groups of definitions see AI in a much broader context, as a science; for example, Stone et al. (2016) say that "Artificial Intelligence (AI) is a science and a set of computational technologies that are inspired by—but typically operate quite differently from—the ways people use their nervous systems and bodies to sense, learn, reason, and take action" (Stone et al., 2016). The Oxford English Dictionary gives this definition: "The theory and development of computer systems able to perform tasks normally requiring human intelligence, such as visual perception, speech recognition, decision-making, and translation between languages". Merriam-Webster Dictionary connects both previously mentioned aspects of Artificial Intelligence and defines it as a) a branch of computer science dealing with the simulation of intelligent behaviour in computers, and b) the capability of a machine to imitate intelligent human behaviour (Merriam Webster, 2018).

In this chapter, we adopted the definition of AI given by Luckin et al. (2016) who define AI as computer systems that have been designed to interact with the world through capabilities (for example, visual perception and speech recognition) and intelligent behaviours (for example, assessing the available information and then taking the most sensible action to achieve a stated goal) that we would think of as essentially human.

These computer systems include a wide range of technologies and methods such as machine learning, adaptive learning, natural language processing, data mining, crowdsourcing, neural networks and algorithms.

1.2 ARTIFICIAL INTELLIGENCE AND HUMAN ACTIVITIES

According to Leontiev's activity theory, human processes can be examined from the perspective of three levels of analysis which are hierarchically linked to each other: activity, actions, and operations. This three-level model of human processes provides a useful entry point for understanding AI and its potential impact on human activities. Societies and economies are reinventing themselves with users of new AI technologies at the forefront. The impact of AI and automation on human activities is twofold: it can either transform them or replace them. Indeed, AI can support or substitute human activity when it comes to operations and actions, the two lowest levels of human processes.

The top of the hierarchy represents the *activity* and its underlying motive: it dictates the meaning of an activity. Any activity is performed through goal-oriented *actions*. The intermediate level represents *actions*, which are essentially ways of solving those problems that need to be solved to accomplish the activity. *Operations*, in turn, implement the actions using available tools; it is the bottom level. In other words, the level of *activity* provides the foundation for ethics. The *actions* implement the cognitive aspect of the activity, and finally, *operations* are the repetitive routine skills to carry out the *actions*. (Heilweil, 2020). Based on human activities we can divide AI into different types.

1.3 TYPES OF AI

Artificial Intelligence can be divided into various types. The main two types of AI categories are based on capabilities and functionally. Figure 1.1 explains the types of AI (Javapoint, 2021).

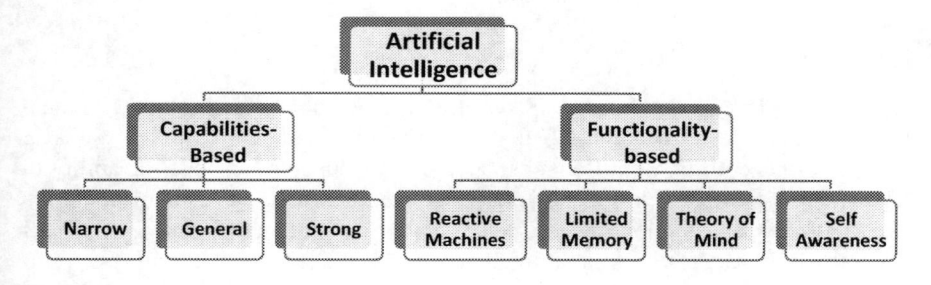

FIGURE 1.1 Types of AI.

1.3.1 NARROW AI

Narrow AI can perform a dedicated task with intelligence. The most common and currently available AI is Narrow AI in the world of Artificial Intelligence. It cannot perform beyond its field or limitations, as it is only trained for one specific task. Hence it is also termed as weak AI. Narrow AI can fail in unpredictable ways if it goes beyond its limits.

1.3.1.1 Examples

Apple Siri operates with a limited pre-defined range of functions. Siri often has problems with tasks outside its breadth of abilities. IBM's Watson supercomputer applies cognitive computing, machine learning, and natural language processing to process information and answer queries. It uses an Expert system approach combined with machine learning and natural language processing. Other examples of Narrow AI include Google Translate, image recognition software, recommendation systems, spam filtering, Google's page-ranking algorithm, playing chess, purchasing suggestions on e-commerce sites, speech recognition, and self-driving cars.

1.3.2 GENERAL AI

General AI is a type of intelligence that could perform any intellectual task with the efficiency that a human being can. It allows a machine to apply knowledge and skills in different contexts. The idea behind General AI was to make a system that could think like a human on its own. Currently, no such system exists which could qualify as general AI, performing any task as perfect as a human. Researchers worldwide are now focused on developing machines with General AI. As systems with General AI are still under research, it will take lots of effort and time to develop such systems.

1.3.2.1 Examples

Fujitsu has built the K computer, which is one of the fastest supercomputers in the world. It is one of the significant attempts at achieving strong AI. It took nearly 40 minutes to simulate a single second of neural activity. Hence, it is difficult to determine whether strong AI will be achieved shortly. Additionally, Tianhe-2 is a supercomputer that was developed by China's National University of Defense Technology. It holds the record for cps (calculations per second) at 33.86 petaflops (quadrillions of cps).

1.3.3 SUPER AI

Super AI is a level of intelligence in systems at which machines could beat human intelligence and perform any task better than humans with cognitive abilities. It is an outcome of general AI. Some key characteristics of Super AI include the ability to think, reason, solve puzzles, make judgments, plan, learn, and communicate on its own. It also evokes emotions, needs, beliefs, and desires of its own. Super AI is still a hypothetical concept of Artificial Intelligence. The development of such systems in reality is still a world-changing task to come.

1.3.4 REACTIVE MACHINES

Reactive machines are the most basic types of Artificial Intelligence. Such AI systems do not store memories or past experiences for future actions. These machines only focus on current scenarios and react to them as per possible best action.

1.3.4.1 Example

IBM's Deep Blue that defeated chess grandmaster Garry Kasparov is a reactive machine that sees the chessboard pieces and reacts to them. Deep Blue cannot refer to any of its prior experiences or improve with practice. It can identify the pieces on a chessboard and know how each moves. Deep Blue can make predictions about what moves might be next for it and its opponent. It ignores everything before the present moment and looks at the chessboard pieces as it stands right now and chooses from possible next moves. Google's AlphaGo is also an example of a reactive machine.

1.3.5 LIMITED MEMORY

Limited memory machines can store past experiences or some data for a short period to make decisions.

1.3.5.1 Example

Self-driving cars are one of the best examples of Limited Memory systems. These cars can store the recent speed of nearby cars, the distance of other cars, speed limit, and other information to navigate the road. Limited Memory AI observes how other vehicles are moving around them, at present, and as time passes. This ongoing, collected data gets added to the AI machine's static data, such as lane markers and traffic lights. They are included when the vehicle decides when to change lanes, avoid cutting off another driver, or hit a nearby vehicle. Mitsubishi Electric has been figuring out how to improve such technology for applications like self-driving cars. Another example is speech recognition technology, like chatbots, which have become a major part of everyday communication and language processing.

1.3.6 THEORY OF MIND

Theory of Mind AI should understand human emotions, people, beliefs, and be able to interact socially like humans. This type of AI machine is still not developed, but researchers are making lots of efforts and improvements for developing such AI machines.

1.3.6.1 Example

Kismet is a robot head made in the late 1990s by a Massachusetts Institute of Technology researcher. Kismet can mimic human emotions and recognize them. Both abilities are key advancements in Theory of Mind AI, but Kismet can't follow gazes or convey attention to humans. Sophia from Hanson Robotics is another example where Theory of Mind AI was implemented. Cameras present in Sophia's eyes, combined with computer algorithms, allow her to see. She can sustain eye contact, recognize individuals, and follow faces.

1.3.7 SELF-AWARENESS

Self-Awareness AI is the future of Artificial Intelligence. These machines will be super intelligent and will have their own consciousness, sentiments, and self-awareness. These machines will be smarter than the human mind. Self-Awareness AI does not exist in reality still and it is a hypothetical concept.

1.3.7.1 Example

One example is a robot "arm" created by a group from Columbia University. The robot learns what it is by itself; it has no prior knowledge, but after a day of "babbling", the robot creates a self-simulation of itself.

1.4 HISTORY OF ARTIFICIAL INTELLIGENCE

Artificial Intelligence is not a new word and not a new technology for researchers. This technology is much older than you would imagine. There are even myths of mechanical men in Ancient Greece and Egypt. To follow are some milestones in the history of AI which mark the journey of its development to date (Javapoint, 2021).

1.4.1 MATURATION OF ARTIFICIAL INTELLIGENCE (1943–1952)

- **1943:** the first work which is now recognized as AI was done by Warren McCulloch and Walter pits in 1943. They proposed a model of **artificial neurons**.
- **1949:** Donald Hebb demonstrated an updating rule for modifying the connection strength between neurons. His rule is now called **Hebbian learning**.
- **1950:** Alan Turing was an English mathematician who pioneered machine learning in 1950. Alan Turing published **"Computing Machinery and Intelligence"** in which he proposed a test that can check a machine's ability to exhibit intelligent behaviour equivalent to human intelligence, known as a **Turing test**.

1.4.2 THE BIRTH OF ARTIFICIAL INTELLIGENCE (1952–1956)

- **1955:** Allen Newell and Herbert A. Simon created the "first Artificial Intelligence program" which was named **"Logic Theorist"**. This programme proved 38 of 52 mathematical theorems and found new and more elegant proofs for some theorems.
- **1956:** The word "Artificial Intelligence" was first adopted by American Computer scientist John McCarthy at the Dartmouth Conference. For the first time, AI was coined as an academic field.

At that time high-level computer languages such as FORTRAN, LISP, or COBOL were invented. Enthusiasm for AI was very high at this time.

1.4.3 THE GOLDEN YEARS-EARLY ENTHUSIASM (1956–1974)

- **1966:** researchers emphasized developing algorithms that can solve mathematical problems. Joseph Weizenbaum created the first chatbot in 1966, which was named ELIZA.
- **1972:** the first intelligent humanoid robot was built in Japan, named WABOT-1.

1.4.4 THE FIRST AI WINTER (1974–1980)

- **1974 to 1980** were like a winter season in the development of AI. An AI winter refers to a period where computer scientists dealt with a severe shortage of funding from the government for AI research.
- During AI winters, an interest in publicity on Artificial Intelligence was decreased.

1.4.5 A BOOM OF AI (1980–1987)

- **1980:** After the AI winter, AI came back with an "Expert System". Expert systems were programmed that emulate the decision-making ability of a human expert.
- **In 1980,** the first national conference of the American Association of Artificial Intelligence was held at Stanford University.

1.4.6 THE SECOND AI WINTER (1987–1993)

- **1987 to 1993** were the second AI Winter period.
- Again, investors and the government stopped funding for AI research due to high costs and inefficient results. However, the expert system XCON was very cost-effective.

1.4.7 THE EMERGENCE OF INTELLIGENT AGENTS (1993–2011)

- **1997:** IBM Deep Blue beat world chess champion Gary Kasparov, and became the first computer to beat a world chess champion.
- **2002:** for the first time, AI entered the home in the form of Roomba, a vacuum cleaner.
- **2006:** AI flooded the business world. Companies like Facebook, Twitter, and Netflix started using AI.

1.4.8 DEEP LEARNING, BIG DATA AND ARTIFICIAL GENERAL INTELLIGENCE (2011–PRESENT)

- **In 2011,** IBM's Watson won the quiz show *Jeopardy!* where it had to solve complex questions as well as riddles. Watson had proved that it could understand natural language and can solve tricky questions quickly.

- **2012:** Google launched an Android app feature "Google Now", which was able to provide information to the user as a prediction.
- **In 2014**, Chatbot "Eugene Goostman" won a competition in the Turing test.
- **2018:** The "Project Debater" from IBM debated on complex topics with two master debaters and also performed extremely well.
- Google demonstrated an AI programme "Duplex", a virtual assistant taking hairdresser appointments on the phone, and the lady on the other end didn't notice that she was talking with a machine.

1.5   AI AND EDUCATION

Teaching how to use computers is not teaching about AI, nor is digital education. AI principles and functions can be taught without the use of IT. There must be some awareness that AI can assist schools as much as it can be taught at school. AI is reforming the fundamental practicalities of education, teaching, and learning. As was pointed out earlier, AI impacts human activity and this applies to teaching methods as well. Machine learning in education may indeed be an opportunity to walk away from a "one size fits all" approach and to develop teaching tools that personalize learning and are tailored to individual needs and capabilities (Karsenti et al., 2017). Rouhiainen (2019) envisions "AI-based learning systems that would be able to give educators useful information about their student's learning styles, abilities, and progress, and provide suggestions for how to customize their teaching methods to students' individual needs". This would allow students experiencing learning difficulties, for instance, to acquire extra tutoring that specifically addresses the identified gap.

As human-machine partnership enters the world of education, curricula will need to be rethought, and teachers and education designers will need to "go meta". In its report on AI in education "Lead the leap", Unesco (2019) claims that education should not be about making the curriculum more technological, but about teaching more human-centric skills—shifting the focus to the top level of Leontiev's pyramid, the *activity*. Unesco's Beijing consensus on AI and education reminds us of the human rights implications of "preparing all people with the appropriate values and skills needed for effective human-machine collaboration in life, learning and work, and for sustainable development". Generally, the importance of promoting these types of values and skills goes unnoticed.

1.6   THE FUTURE OF ARTIFICIAL INTELLIGENCE IN EDUCATION

The pandemic came suddenly and unpredictably for the whole world, especially for teachers and students. Engaging students became very difficult when all you have to work with is a screen, and with little skill. Everyone was confused about where and how to start. Therein lies the future of AI. The education system of the future will probably engage Artificial Intelligence to improve recognition and personalized learning for every single student. Teachers, in the past, employed a "one size fits all" approach to education. Due to lack of resources and the need to meet specific learning targets by set dates, it was necessary to teach to the middle. This overlooked the underperforming students who were slower to grasp the concepts. Artificial

Intelligence software that learns and adapts to each student's individual needs, however, offers the opportunity for teachers to help exactly where they are needed and it will tailor the learning experience so that no student gets left behind. This is a truly personalized approach to learning. All teachers must be equipped with this skill.[1]

Automation, on the other hand, focuses on making dull tasks easy for both students and teachers. Students can benefit from using AI technology to make some learning tasks easier and more interesting. Translating languages in real-time, for example, makes knowledge more accessible for students globally. Teachers can benefit from automation with the introduction of automatic grading software that will free up valuable time currently spent marking student papers. Intelligent teaching assistants that can automate some question and answer tasks teachers perform will allow students to access help whenever they need it without soaking up valuable teacher time.

The advantage of using Artificial Intelligence in a learning environment is its capability to familiarize with changing environments. It can help improve student engagement during remote learning and even provide additional study-aid material. In the future, education will be faster paced and educational needs much more varied. Recognizing trends before they take hold and adapting quickly is an area in which Artificial Intelligence can be of huge help. The educational institutions of the future will be able to change their curricula easily when necessary.

The future looks bright for the potential of machine learning and AI to transform education, though its utilization will vary all over the world. We may notice improvements already underway. In both rural and urban areas, the world of education will be very technologically diverse a few decades from now.

1.7 ROLES OF AI IN EDUCATION

To advance the study process we may apply the following AI tools (Lisa Plitnichenko, 2020):

1.7.1 CUSTOMIZE EDUCATION

Artificial Intelligence helps determine what a student does and doesn't know, allowing one to build a customized study schedule for each learner in consideration of their knowledge gaps. In such, AI can tailors studies to a student's specific needs, increasing their efficiency. Many companies train their AIs, armed by the knowledge space theory, to identify knowledge gaps, taking under consideration the complexity of scientific concepts relative to one another (for instance, one can stimulate the training of another or become a basis for remedial training).

1.7.2 PRODUCE SMART CONTENT

- Digital lessons
  Digital learning interfaces with customization options, digital textbooks, study guides, bite-sized lessons, often generated with the assistance of AI.

---

1 www.rev.com/blog/the-role-of-artificial-intelligence-in-education

- Information visualization
  New ways of portraying information like visualization, simulation, web-based study environments—often powered by AI.
- Learning content updates
  AI helps generate, manage, and update learning material, as well as customize it for various learning curves.

1.7.3 CONTRIBUTE TO TASK AUTOMATION

Reviewing, evaluating, and answering students is a tedious task that is enhanced by the teacher utilizing AI. These days the clues Gmail gives in the messages you write depend on the outline of your current and past messages in addition to the business jargon basics. It is extraordinary to have such an option on any Learning Management System. Teachers may delegate a set of monotonous tasks to AI, allowing them to focus on advancing the quality of the lessons.

1.7.4 ORGANIZE TUTORING

Teachers must constantly develop individual educational programs that take into account student's gaps. Personal tutoring and support for students outside of the classroom assist learners in keeping up with the course, so that parents are not struggling to explain algebra/geometry and statistics to their kids. AI tutors are great time-savers for teachers, as they do not have to invest additional energy explaining challenging topics to students. With AI-powered chatbots and AI virtual individual collaborators, students can avoid being embarrassed by requesting extra assistance in front of their classmates.

1.7.5 ENSURE ACCESS TO EDUCATION FOR STUDENTS WITH SPECIAL NEEDS

The implementation of advanced AI technologies opens up better approaches to cooperate with students with learning disabilities. AI grants access to education for students with special needs such as people who are deaf and hard of hearing, visually impaired, and people with ASD. Artificial Intelligence tools can be trained to assist any group of students with special needs.

1.8 LEARNING

Artificial Intelligence (AI) makes it conceivable for machines to gain knowledge, amend new contributions, and perform human-like tasks. Among many fields, AI blends well in education and can be applied to learning. AI's digital, dynamic nature also offers opportunities for student engagement that cannot be found in often regularly obsolete documents or fixed environments. Synergistically, AI has the potential to boost and accelerate the discovery of new learning limits and the creation of innovative technologies.

Though yet to become standard tool in all organizations and schools, Artificial Intelligence in learning or training has been a "big thing" since AI's uptick in the 1940s (when the first seeds of AI were sown with programmable computers). From various perspectives, the two appear to be made for one another. A new report from eSchool News found that the utilization of AI in the schooling business will develop by 47.5% through 2021. The technology will impact everywhere from kindergarten through higher education to corporate training, offering opportunities to create adaptive learning features with customized devices to improve learners' experiences (Kuang, 2017).

1.9 EXAMPLES OF ARTIFICIAL INTELLIGENCE IN LEARNING

Here are four ways AI is changing the learning industry (Suresh, 2018):

1.9.1 SMART LEARNING CONTENT

The notion of brilliant content is a trendy topic now as AI can create digital content with a similar level of grammatical ability as their human counterparts. Smart learning content creation, from digitized guides of textbooks to customizable learning digital interfaces, are being introduced at all levels, from elementary to post-secondary to corporate environments. One of the approaches to utilize this is when AI can consolidate the content in inconveniencing troubleshooting guides into more acceptable and easy study guides with troubleshooting steps, summaries, flashcards, and intelligent simulations. Smart learning content can also be used to design digital curricula and content across a variety of devices, including video, audio, and online assistants.

1.9.2 INTELLIGENT TUTORING SYSTEMS

AI can do more than condense a lecture into flashcards and smart study guides as it can also tutor a learner based on the difficulties they're having. This involves something known as "Mastery Learning". Mastery learning is a set of principles largely tied to the work of educational psychologist Benjamin Bloom in the 1970s. This supports the effectiveness of individualized tutoring and instruction in the classroom. There are now smart tutoring systems that use data from specific learners to give them feedback and work with them directly. For instance, an Intelligent Tutoring system called "SHERLOCK" is being used to teach Airforce technicians to diagnose electrical system problems in aircraft. Another advanced version of Intelligent Tutoring Systems is avatar-based training modules which were developed by the University of Southern California to train military personnel being sent on international posts. While this AI application is still in its early stages, it will soon be able to work as a full-fledged digital platform that helps learners with their educational needs in just about any area. These platforms will soon be able to adapt to a wide variety of learning styles to help every educator and learner.

1.9.3 VIRTUAL FACILITATORS AND LEARNING ENVIRONMENTS

With AI, an actual lecturer may soon be replaced by a robot. Well, not entirely! But there are already virtual human mentors and facilitators that can think and act like humans. But how does a virtual facilitator think or act like a human?

A new trending technology known as "touchless technology" or "gesture recognition technology" gives virtual facilitators the ability to respond or act like humans in a natural way, responding to both verbal and nonverbal cues.

Smart learning environments and platforms use AI, 3-D gaming, and computer animation to create realistic virtual characters and social interactions. This initiative includes more than virtual facilitators, as Augmented Reality may soon be a part of the training.

1.9.4 CONTENT ANALYTICS

Content analytics refers to AI (specifically machine learning) platforms that optimize learning modules. Through AI, content taught to learners can be analyzed for maximum effect and optimized to take care of learner needs. Content analytics enables educators and content providers to not just create and manage their eLearning content, but also gain important insights into learner progress and understanding through a powerful set of analytics.

1.10   AI IN THE CLASSROOMS

A very common question related with AI in the classroom is whether robots will replace professors, whether the effects of progress will be positive or negative, and what should be done to improve current teaching approaches. Apprehension surrounded the possibility of using Artificial Intelligence for educational purposes until a series of helpful digital tools were tested in schools and universities. Optimists believe that it is the beginning of a great tech era, a step towards fantastic growth and progress. Simultaneously, pessimists are claiming that machines can go out of control and lead humanity to a state similar to what was depicted in futuristic films like "The Terminator". As controversy heats up, optimists invest their funds in creating new AI tools, adaptive programs, and smart robots. There are several options for using innovative technologies in the classroom (Kerry Rose, 2020), (Vinichenko, M. V., Narrainen, G. S., Melnichuk, A. V., & Chalid, P., 2021).

1.10.1 TUTORING

While machines cannot replace a live teacher, they can greatly simplify one's work. It is impossible to work individually with each learner in a class of more that 20 learners, answer everyone's questions, and give personal development recommendations. With AI, students may receive extra teaching time. Computers are ready to support them and provide necessary information. This helps students struggling with challenging topics to study at a comfortable pace and eliminate knowledge gaps. Teachers can't be with their students 100% of the time, and that's why tutors are important. Some students just need a little bit of extra help with a certain subject matter. For many introverted people, seeing a tutor can be daunting, and going to a tutoring lab can be even more unnerving. Working with an AI Tutor can help students with social or academic anxiety get the help and confidence they need to succeed in school

1.10.2 GRADING

Checking homework and control papers are one of the most unloved aspects of teachers' work. Fortunately, they can delegate these responsibilities to AI and spend more time communicating with students, studying educational trends, and developing advanced innovative teaching approaches. Computer tests are used in many educational institutions all over the world, especially for competitive entrance tests. This kind of system is not new; I remember when we sat for medical entrance test back in 1988, this system was in place, counting correct answers and showing marks instantly. In addition, there are softwares to check essays, detecting grammar and punctuation mistakes. Though there are still no programs able to assess creativity and emotional component of a text, they may be created soon.

1.10.3 SIMPLIFYING TRIAL-AND-ERROR

Mistakes are important because they teach us to do things correctly. However, failure to answer a question in front of classmates and teachers may be rather frustrating, lowering self-esteem and learning enthusiasm. AI is a great secret-keeper. A student may not be afraid to receive harsh criticism. One is free to experiment in the digital space, receive useful recommendations from a smart assistant, and solve problems much faster than in real life. In this way we can say AI can be a confidence booster.

1.10.4 VIRTUAL REALITY

Virtual reality (VR) will be an integral part of the next generation's educational experience. Imagine being able to transport students to ancient Rome and the Colosseum to learn about gladiators or to the Sahara Desert to teach kids about different climates or animals—all without spending a penny on travel or requiring teachers to be completely knowledgeable about the subject matter. VR expands the boundaries of a physical classroom. It allows learners to explore amazing universes, conduct complex experiments, and observe processes that would otherwise be too fast or slow. The learning process becomes more fun, engaging, and demonstrative. Young people may not only read about ancient civilizations and celestial bodies but also see them, acquire an unforgettable experience without leaving a comfortable classroom, and develop skills in disciplines that were previously taught only theoretically. Also, virtual trips may replace educational tourism, helping travelers to save much time and money. Immersive learning goes hand in hand with AI-assisted education, and VR has already increased students' abilities to become completely immersed in a lesson. Price is still the biggest roadblock for VR and the educational system, but the longer these technologies are around, the cheaper they'll become—this same thing happened with computers and phones, and it'll likely happen with VR as the technology is integrated more into classrooms.

1.10.5 GLOBAL LEARNING

Artificial Intelligence (AI) is making headlines in newsrooms across the country. One of the latest trends we've seen is in the education system, and it's made many people suspicious of the consequences that using Artificial Intelligence in the classroom will have. Information from the whole world is at students' fingertips. They can consult famous researchers, attend online lectures of the most prominent scientists and entrepreneurs, and communicate with peers from other countries, thus strengthening international cooperation and cultural exchange and learning foreign languages from native speakers. We must not take this as a threat. This is going to make all the teachers more smarter.

1.10.6 GRADING

Electronic grading has existed for many years in the form of computer tests and Scranton testing. Advancements in AI could allow teachers delegate all assignments to an AI for grading so the teachers can spend more time with students individually as well as modify the curriculum to be more relevant for each class.

There are downsides to having an AI grade papers, though. It does require a constant internet connection to work effectively, and the AI itself is expensive. However, "the time saved in the classroom with an AI will ultimately pay for itself time and time again", states Joshua (2021) Adamson-Pickett of Business.org. Despite the cost and need for internet, AI makes for a phenomenal grading tool.

1.10.7 ANALYTICS (CURRICULUM, TEACHING STYLE, ETC.)

Being able to adjust curricula for individual classes helps teachers create an incredible advantage for students, especially because the many different learning and teaching styles don't always work cohesively. An AI teaching assistant would be able to adjust the aforementioned factors, ensuring students can receive the most effective teaching methods for their learning style within a limited time.

1.10.8 TRIAL AND ERROR

Trial and error can be one of the most intimidating and exhausting parts of education. Taking this stage out of teaching and learning can save time and frustration levels for both students and teachers. An AI could take all the time-consuming problems in a teaching-learning setting and solve these troubles instantly, freeing up the teachers' and students' time for other aspects of education. An AI could also explain how it solves problems to help educators and students mitigate the same situation in the future.

1.11 EVOLUTION OF EDUCATION THROUGH AI

Technology's impact on the world of education strengthens with each year. It is clear that great changes are coming, and machines will play a direct role in them. Schools and universities will never return to the original format. AI has brought wonderful transitions in the education sector, benefiting both the students and the institutions. Let's see how AI is changing the education industry.

Artificial Intelligence in education is growing, and outdated methods are changing radically. Yes, Covid-19 played vital role in that. The academic world is becoming more convenient and personalized due to the numerous applications of AI for education. This has changed the way people learn since educational materials are becoming accessible to all through smart devices and computers. Today, students don't need to attend physical classes to study as long as they have computers and an internet connection. AI is also enabling the automation of administrative tasks, allowing institutions to minimize the time required to complete difficult tasks so that the educators can spend more time with students. Now is the time to discuss the transformations brought by AI in education. Let's have a detailed review:

1.11.1 Simplifying Administrative Tasks

AI can automate the expedition of administrative duties for teachers and academic institutions. Educators spend a lot of time on grading exams, assessing homework, and providing valuable responses to their students. But technology can be used to automate grading tasks where multiple tests are involved. This means that professors have more time with their students rather than spending long hours grading them. We expect more of this from AI. Actually, software providers are coming up with better ways of grading written answers and normal essays. Another department gaining a lot from AI is school admissions boards. Artificial Intelligence is allowing for automation of classification and processing of paperwork.

1.11.2 Smart Content

AI and education go hand in hand and new technologies could significantly assist all students in attaining their ultimate academic success. Smart content is a very hot topic today. Robots can produce digital content of comparable quality as that which essay writing services create. This technology has already reached a classroom setting. Smart content also includes virtual content like video conferencing and video lectures. As you can imagine, textbooks are taking a new turn. AI systems are using traditional syllabuses to create customized textbooks for certain subjects. As a result, textbooks are being digitized and new learning interfaces are being created to help students of all academic grades and ages. An example of such mechanisms is Cram101, which uses AI to make textbook contents more comprehensible, and it is easy to navigate with summaries of the chapters, flashcards, and practical tests. The other useful AI interface is called Netex Learning, which enables professors to create electronic curricula and educational information across myriad devices. Netex includes online assistance programs, audios, and illustrative videos.

1.11.3 Personalized Learning

The traditional systems are supposed to cater to the middle but don't serve all pupils sufficiently. The curriculum is designed to suit as many pupils as possible by targeting 80% of the middle. However, some pupils struggle to attain their full potential when in the top 10%, while those in the bottom 10% have difficulties following along. When AI is introduced, teachers are not necessarily replaced, but they are in a position to perform much better by offering personalized recommendations to each pupil. AI customizes in-class assignments as well as final exams, ensuring that students get the best possible assistance.

Research indicates that instant feedback is one of the keys to successful tutoring. Through AI-powered apps, students get targeted and customized responses from their teachers. Teachers can condense lessons into smart study guides and flashcards. They can also teach students depending on the challenges they face in studying class materials. Unlike in the past, college students can now access a larger window time for interacting with professors. Thanks to AI, smart tutoring systems, like Carnegie Learning, can offer quick feedback and work directly with students. Even though these methods are still in their inception stages, they will soon become fully-fledged digital teachers to assist students with any educational needs.

1.11.4 Global Learning

Education has no limits, and AI can help to eliminate boundaries. Technology brings drastic transitions by facilitating the learning of any course from anywhere across the globe and at any time. AI-powered education equips students with fundamental IT skills. With more inventions, there will be a wider range of courses available online and with the help of AI, students will be learning from wherever they are.

1.11.5 New Efficiencies

AI improves IT processes and unleashes new efficiencies. For instance, town planners could use it to minimize traffic jams and improve the safety of pedestrians. Similarly, schools can determine the appropriate methods of preventing students from getting lost in crowds when they run in corridors. AI can also be used in the modeling of complex data to enable the operations department to create data-driven forecasts. This, in turn, allows proper planning for the future; for example, assigning seats during school functions or ordering food from local cafeterias. Speaking of which, schools can avoid a lot of wastages caused by over-ordering thereby saving costs. Through new efficiencies, Artificial Intelligence in education can pay for itself. The truth is new technologies come with upfront expenses for installation and training. But eventually, these costs become negligible. Technology gets cheaper over time and so does the hardware and software.

A study published by eSchool News indicates that by 2021, the application of AI in education and learning will be increased by 47.5%. The impact of this technology will be felt from the lowest education levels through higher learning institutions. This will create adaptive learning techniques with customized tools for improving learning experiences. Artificial Intelligence could inform the students how their career paths look like depending on their goals thus assisting them beyond academics. Only time can tell the ultimate impact of AI in the education industry.

1.12   WHAT KIND OF SCHOOLS DO WE WANT IN THE FUTURE?

Is there a need to prepare teachers-in-training to work with AI? Yes—for so many reasons. AI already has a heavy influence on individuals and societies, and we need to develop critical perspectives on AI issues. If teachers are trained in AI use, it will help prevent technology abuses. Importantly, for AI to make a real contribution to academic success, and for all students, the teacher's role remains as central as ever, perhaps more than ever. Because intelligent robots will transform tomorrow's workplaces, children should begin preparing for the new reality as early as primary school. The technology players must not be allowed to have the only say in all this. We have seen that AI has penetrated all the education spheres, in the form of intelligent books, web browsers, education apps, and learning platforms, to name a few. Yet "education" gets only one mention on the Wikipedia page devoted to "Artificial Intelligence". The question therefore arises: what is our vision for schools of the future? Will the technology giants be solely in charge of the ways that Artificial Intelligence is used for learning? Or will students and teachers be able to ask questions and provide clear, constructive, responsible, and ethical guidelines for how technology is used? The money-makers should not be allowed to decide our future without our say-so. The uses of AI must be carefully planned by the entire spectrum of education actors, starting with the teachers and learners. All this aligns with Québec's vision for its proposed international observatory on the societal impacts of Artificial Intelligence and digital technologies. According to Rémi Quirion, Chief Scientist of Québec, it will serve as an international hub for AI research, and it will demonstrate Québec's leadership in the ethically and socially responsible development of AI. Many experts, including Yoshua Bengio at the Université de Montréal, worry about the risks of AI for society if the forward march of intelligent machines is left unchecked. But before we apply the breaks, we need to have some idea of how it works and what it does. It would also be wise to bring in advisors and experts who are not technology stakeholders. From now on, we'll have to live with AI, so isn't it about time that we get our teachers ready for it?

In terms of the benefits for education, AI is neither a panacea nor the Holy Grail. Instead, it's a tool with tremendous potential, and one that we must learn to use to its full advantage. One major challenge is to find the right balance between the time-honoured teaching practices that have been handed down for centuries and the new possibilities afforded by AI. Moreover, we should not limit ourselves to a simple utilitarian vision of AI. Instead, we could try to imagine how learning can be nurtured and transformed. Ultimately, AI is not just good for school: it's also good for creating greater understanding and respect between all people

The potential of AI in education has been researched, debated, and discussed for nearly 30 years within the AIED academic community. Over the last few years, the debate has crept into the international public policy arena as data, sophisticated AI algorithms, computing power, and access to technology has increased across the world. There are great potential benefits, but there are risks as well. We therefore need to proceed diligently and prudently into a new educational environment where AI is used to support learners and teachers, and where we also prepare learners for a future world in which AI plays an increasing role.

The risk is that the education system will be churning out humans who are no more than second-rate computers, so if the focus of education continues to be on transferring explicit knowledge across the generations, we will be in trouble. The AI challenge is not just about educating more AI and computer experts, although that is important. It is also about building skills that AI cannot emulate. These are essential human skills such as teamwork, leadership, listening, staying positive, dealing with people and managing crises and conflict.

Financial Times (2017)

New technology is evolving every day in this fast-moving, data-driven age. Even less than two decades ago, smartphones were just emerging and were capable of functioning the rudimentary phone operations only. Since then, technology has taken a huge leap—from shopping to watching a movie, everything can be possible on a smartphone. So, the traditional methods of functioning in our lives are changing and with that, one of the most important sectors to consider—education sector—should also adapt to these changes.

Applying new evolving technologies like Artificial Intelligence (AI) can change the whole system and lead to the emergence of a more transparent and efficient education system. With world literacy rate as high as 86.3% and subsequent expansion in the field of education, the issue of implementation of AI in this sector becomes important.

1.13   WHAT IS THE NEED FOR AI IN EDUCATION?

While our way of living and daily operations has been rapidly altered, to date, traditional methods are generally still practiced in the education sector. Kids today are mostly not engaged in traditional games and toys and colouring books. Most of their activities revolve around laptop screens, smartphones, and tablets. Yet, today's education systems stay still more static and traditional. So, a change in the paradigm of education is needed to keep up with the changing requirements and lifestyles of the students.

Developing countries like India, Swaziland, and South Africa are still waking up to the new concepts of digitization, and the education industry being still untouched will not only benefit the student when it comes to harnessing their individualistic potential, but will also allow the trainers some space by letting go off some mundane tasks that they do in today's traditional way of teaching.

Mr. Abhishek Manjrekar, CMO—Marketing & Analytics at
Foreign Academic Consultancy & Training (FACT)

1.14   WAYS IN WHICH AI CAN IMPROVE EDUCATION

There are various areas for application of AI in education, some of which are discussed as follows:

With the application of AI, basic activities in education, like grading, can be automated. Grading can be a tedious task and it takes up a lot of time of the professors.

So, automated grading will save time for teachers so that they can use this significant amount of time to interact with students.

Learning can be individually customized for a student based on his or her needs, requirements and pace. Educational software can be used accordingly. With this system of adaptive learning, students will receive individual care which is not feasible in traditional educational processes. Moreover, each student can be guided by a virtual mentor in case of queries.

According to industry experts, education with AI can assist learners to gain and address the required skills of the 21st century like self-assessment, confidence, teamwork, and others more efficiently.

In most of the traditional teaching practices, teachers are often not aware of the shortcomings of their lectures and thus there is always a gap in courses and learning. With large numbers of students in a course submitting various queries and wrong assignments, it becomes humanly impossible for a mentor to address all queries. AI can solve this problem. One of the biggest online course providers, Coursera, is already starting to apply this technology for their courses. AI can also help to monitor the progress of the students and report professors if there is any issue with the student performances.

AI can change the role of teachers. With AI, teachers will have to adapt themselves to be concerned with learning issues only instead of manual tedious works like grading. AI can also create a platform for students to find information and address queries. One of the most common issues a teacher faces while teaching a group of students is the pace of the course. With AI, this problem will be virtually solved as each learner can learn at their comfortable pace. So teachers will supplement the AI lessons to assist students while they provide human interaction.

The technology will also provide opportunities for a global classroom with increased accessibility and participation of students worldwide. Without technology, it would be difficult to maintain a platform of this scale. It opens up educational possibilities to each part of the world and is a highly cost-saving procedure.

1.15 CONCERNS OVER THE ISSUE

It is anticipated that with more use of AI in education, the role of teachers could become redundant. However, advocates of this technology suggest that AI should be viewed as a technology to support teachers and not to replace them.

If AI benefits only students with access to specific advanced technologies, then it may marginalize some groups. It is also debated that the students might become too dependent on the technology. These things must be analyzed before implementing AI on a large scale. "AI being an automated and real-time system may threaten to cease the personal bond between a trainer and student when it would come to virtual classes or offshore assignments", adds Mr. Abhishek Manjrekar.

Any technology costs big money for initial investment. Given the Indian scenario, where many institutions fail to meet even the basic educational standards amidst a plethora of institutions, setting up an AI system will not only incur huge costs but also require trained manpower who can understand it fully and implement it with effective use with the maximum effect, Manjrekar further added.

Moreover, educators will also have to be careful regarding privacy and confidentiality of the data collected and its access.

1.16   HOW TO START IMPLEMENTING AI

If you're considering AI as an option to customize the learning experience, these steps will help you to plan your project.

1. **Identify your needs and AI technologies**
   The starting point of implementing any technology is the identification of the pain points this technology can address and resolve. Find the system bottlenecks and research the ways AI could optimize these processes.
2. **Determine the strategic objectives of AI transformation in your organization**
   Determine your appetite: do you want to be an early adopter or a follower? Which technologies will fit your company best? Are you aware of the AI drawbacks and how are you going to address them? The business objectives to which AI technology should contribute? Based on responses to these questions, you should develop a cost-benefit analysis for AI automation and augmentation.
3. **Make the right culture, talent, and technology meet**
   To make the most of the AI tools, you should not only choose the right team to adopt the technology but also create the right environment driven by analytical insights and focused on actionable decisions on all organizational levels.
4. **Smart ways to control the outcome of AI transformation**
   Creating an environment for both human beings and AI to work side-by-side, it's important to ensure the processes' transparency and keep pace with the key considerations and metrics of AI adoption. Based on the custom characteristics of your organization and type of AI implemented, decide on the performance indicators to track, security concerns to keep under control, and technical ecosystems to support.

1.17   ADVANTAGES OF ARTIFICIAL INTELLIGENCE

To follow are some of the main advantages of Artificial Intelligence:

- **High Accuracy with fewer errors:** AI machines or systems are prone to less errors and high accuracy as it takes decisions as per pre-experience or information.
- **High-Speed:** AI systems can perform fast decision-making at very high-speeds; because of this, AI systems can beat a chess champion.
- **High reliability:** AI machines are highly reliable and can perform the same action multiple times with high accuracy.
- **Useful for risky areas:** AI machines can be helpful in situations where employing a human can be risky, such as defusing a bomb or exploring the ocean floor.

- **Digital Assistant:** AI can be very useful to act as digital assistants to users, such as AI technology currently used by various e-commerce websites to show the products as per customer requirement.
- **Useful as a public utility:** AI can be very useful for public utilities such as a self-driving cars which can make our journey safer and hassle-free, facial recognition for security purpose, natural language processing, and more.
- **Quick Feedback:** automatic correction of certain kinds of schoolwork, which frees up teachers' time for other tasks. Unfortunately, the current thesis correction software (mostly available in English) leaves something to be desired. Despite the amazing progress made, the human touch remains essential.
- **Ongoing student assessment**: learners' experiences along the learning pathway are tracked in real time to accurately gauge skills acquisition over time.
- **Intelligent tutoring platforms** for distance learning: this is a growing trend, and combined with the rapid expansion of mobile technology, it opens up exciting opportunities for learners and educators alike.
- **Quick interaction:** new ways to interact with information. For example, Google adjusts our search results according to our geographic location or previous searches, generally without our knowledge. Amazon does the same when it suggests purchases in light of what we bought in the past. Siri, Apple's voice recognition assistant, adapts to individual voices, needs, and requests.
- **Educational feedback**: for example, software can send personalized texts to students as they follow their learning pathway. Not only is the feedback personalized, it's faster and more frequent, it allows automated grading, and it offers support and tailored recommendations.
- **Adapted teaching content**: such as the digital bookshelves published by Pearson and McGrawHill.
- **Communication opportunities**: expanded opportunities for learners to communicate and collaborate with each other. Greater interaction between learners and academic content. An example is the chatbot, an offspring of the original smart speakers like HomePod, Amazon Echo, and Google Home. A chatbot can recognize the user's language and simulate a real conversation.
- **Content facilitation**: better teaching through facilitation rather than content transmission. But make no mistake: the teacher remains the star of the classroom, while AI plays a supporting role by handling complex digital tasks.
- **Homework assistance**: students can do personalized homework that suits their academic skills and challenges. The online homework helper Allô prof, which has assisted students for over 20 years, would certainly benefit from AI.
- **Additional learning**: because AI can personalize exercises to make learning more meaningful and enjoyable.
- **Virtual reality**: these highly interactive, three-dimensional virtual worlds encourage students to engage with course material. For example, the educational game Assassin's Creed lets students appreciate history as they "live" through vivid and detailed historical situations and carry out intriguing missions. Such enriched, interactive experiences have direct positive impacts on learning.

- **Decreased dropout**: AI can gather student data and rapidly warn schools about those who are at risk for dropping out so they can receive appropriate support before matters deteriorate.
- **Accessible**: AI makes distance learning more accessible and appealing. People can learn anywhere, anytime, and programs can be made to measure. The language learning system Duolingo is an outstanding example.
- **Better classroom management**: if the lesson is organized, for example, a virtual experience like Classcraft engages students.
- **Efficient administrative management**: it's not only learners and teachers who benefit but administrations also become more efficient. For example, newsletters, student absences, and so on can be handled quickly and easily.
- **Data collection, storage, and security**: AI's cloud technology allows capturing, organizing, analyzing, and producing knowledge from vast amounts of data, while keeping them secure. This addresses both ethical and educational issues.
- **Automated tasks**: much time that is normally spent on important education tasks can be taken over by AI systems.
- **Humanoid robots**: although they will probably never replace real-life teachers, despite Hollywood fantasies, life-like robots will play an ever larger role in classrooms. They will act as teacher's assistants by performing complex and time-consuming tasks.

1.18   DISADVANTAGES OF EDUCATION AI

However, when you look at the bright side of a thing, you should acknowledge that there is also a darker side to it. Despite several advantages that AI offers, it also has some disadvantages that we can't ignore. Let us look at some of the major disadvantages of AI implementation. Of course, there are no ideal approaches and technologies. It is worth paying attention to the following factors could limit AI use in the academic world.

- **High cost of implementation:** The hardware and software requirement of AI is very costly as it requires lots of maintenance to meet current requirements. Do you know how much it cost Apple to acquire its virtual assistant? The acquisition of the software cost somewhere around a whopping $200 million. Further, the high cost of AI implementation is evident from the fact that Amazon acquired Alexa for $26 million in 2013. These AI-based software programs require frequent upgrades in order to cater to the requirements of the changing environment as the machine needs to become smarter by the day. In case the software suffers a severe breakdown, then the process of recovering lost codes and reinstalling the system can give you nightmares due to the huge time and cost involved.
- **Can't think out of the box:** Even as we develop smarter machines with AI, they still cannot work out of the box, as the robots will only do that work for which they are trained, or programmed.
- **No feelings and emotions:** Though AI machines can be outstanding performers, they still do not have feelings and cannot make any kind of

emotional attachment with humans, and may thus be harmful to users if proper care is not taken.

- **Increase dependency on machines:** With the incremental spread of technology, people are getting more dependent on devices and hence they are losing their mental capabilities.
- **No original creativity:** AI machines cannot beat the power of human intelligence in terms of imagination and creativity.
- **Social Skills:** The pleasure of live communication is essential for creating a favorable emotional atmosphere in a classroom. If interacting only with robots, students will not acquire social skills. Without knowing how to build human connections, a person is likely to face challenges in personal and labor life, feel lonely and unhappy.
- **Infrastructure:** AI requires good infrastructure: the latest computers and a stable Internet connection. Schools with poor funding are unlikely to afford the necessary equipment. Many of them even do not have local Wi-Fi. However, all this is just a matter of time. When a printed book was a great rarity, only churches and wealthy families had a library. Artificial Intelligence is also going to turn into a common thing over some time. Prices will lower as technology spreads.
- **Technical Issues:** No one is immune from technical problems. Computers are subject to virus attacks, can break or freeze at the most inconvenient moment. Large data volumes may be lost because of errors, and it will be rather challenging to restore them. Even though there are many disadvantages, there will be always room for improvement. We can't simply discard it thinking only of its disadvantages.

1.19 IMPACT OF COVID-19 IN FOSTERING AI IN EDUCATION SYSTEM

Around the globe, the Covid-19 pandemic has affected the lives of educators as well as students, socially, economically and culturally. The functioning of the school is another story. Educational institutions were closed and students were no more attending face-to-face classes. These challenging conditions brought an opportunity to reshape old models of higher education. This is the time for transition by developing, implementing, and diffusing AI or digital technologies among academics and students. Covid-19 arrived when higher education needed improvement for sustainable development by redefining its teaching and learning methods, leadership models, and interaction channels; by going digital.

Additionally, educational institutions had to move fast from face-to-face programs to online alternatives, changing the old ways of teaching carried out for many eras before this pandemic. This is the "new normal" (an expression used more and more these days) and the main avenue in facing and successfully coping with the novelty, the fear, and the uncertainty caused by the pandemic (Tesar, 2020). The teaching and learning activities are now much more developed via the Internet, and HEIs are using their specific Internet platforms (e.g., Moodle) or other online tools, such as Zoom or Google Classroom, Google Meet, and Dojo, and e-learning has often become the only possible mode of formal learning [Serpa, 2020].

A recent published study, using a sample of 30,383 students from 62 countries, concluded that these students are overall satisfied both with the transition from face-to-face to online teaching and the support they received from their teachers throughout this process (Aristovnik et al., 2020), (Kahn, K., & Winters, N., 2018). On the other hand, students mentioned the lack of digital competences and the perceived higher workload. Moreover, the study concluded that the students most affected by the Covid-19-related educational changes were male, part-time, undergraduate, and applied sciences students, as well as students with a lower living standard (i.e., students that can only afford their educational costs with the help of a scholarship, and also part-time students that lost their job as a consequence of the pandemic). In addition, according to Aristovnik et al., this scenario is much worse in less-developed regions and countries of Africa and Asia, which stresses the importance of strengthening the educational offerings in these regions to prevent disparities at the digital, social, economic, and gender levels.

In my opinion, these shortages may be attained by taking the following measures:

1. Transition in syllabus is required.
2. AI must be included in teachers training courses to train the teachers well and to enable them to adapt the system by themselves and help students in adapting to the new system.
3. Governments must give full support to enhancing infrastructure and other facilities like teaching and learning tools to students as well as teachers to bring digital literacy.
4. Teaching methods must be restructured.
5. Curricula must also be revised to become more flexible.

I like the strategies proposed by Cheema to help improve this new way of learning and teaching **(Table 1.1).**

1.20   CONCLUSION

Higher educational instutions, academics and students must adapt to change, so as to develop a future-forward educational process. Overall educational institutions, and in particular HEIs, need to reinvent themselves and reshape the learning and teaching process. This complex, yet urgent need encompasses, according to Darling-Hammond et al. (2020), 10 key areas that HEIs have to address so as to achieve transformational, equity-oriented, and quality learning (Figure 1.2).

In the current challenging scenario caused by the Covid-19 pandemic, the answer may be that the whole educational system, and specifically the higher educational system, needs to engage in and commit to a transformational process. Education must question its own role in these troubled times, notably in promoting a fair, equitable, and sustainable society for all. It is the role of education to permanently question the whole notion of sustainable development as the right path to follow, instilling in its students the will to "create new visions and paradigms to make this world a

TABLE 1.1

Strategies and Description given by Source: Cheema (2020).

Strategies	Description
1. Be prepared (preparedness planning and contingency planning)	• Prepare content before class and inform students in advance. Preferable time for each online session is about 30 ± 10 min. • Check capability of institution's online platform to host large scale users. If not, find alternative online platforms (e.g., Google Classroom, Microsoft Teams, Zoom, Webex, etc.). • Check educators' and students' bandwidth to do a live synchronous online session. If bandwidth is low, do asynchronous online learning. • Have information technology/support team on standby in case of technical glitches. • Provide pre-class reading materials to students to ensure engagement and in-depth discussion. • Gauge online learning behaviour characteristics of students.
2. Bite-sized information is gold (dividing teaching content into smaller units)	• This is to increase student attentiveness and concentration during online learning. • Break down the content of the in-class teaching into different topics and adopt a modular teaching method. Ideally each online session addresses one learning outcome. If learning outcomes are huge, break them into smaller chunks. • Interlay quizzes, discussion, games within each module if possible.
3. Personalize/ Humanize (use of "voice" in teaching)	• Body language and facial expressions are restricted in online learning. • Personalize your online teaching by creating your own videos with voices. Creating animation, digital stories are other options. • Educators should appropriately slow down their speech to allow students to capture key knowledge points. • Use interactive teaching pedagogy where possible.
4. Teamwork makes the dream work (working with teaching assistants, technologists, and online support team)	• Educators are insufficiently trained or supported to operate online learning platforms. • Educators should communicate with teaching technologist/information technologist support team prior and during each online session and prepare contingency plan. • Employ "teaching assistants". Make them aware of the objectives, knowledge framework, and teaching activities of each session.
5. Empowerment (strengthening students' active learning ability outside of class)	• Educator has less control over student engagement and participation during online learning, and students are more likely to skip classes. • Educator should use various methods to modify students' homework, activities, and reading requirements to strengthen students' active learning outside of class.
6. Flexibility (combining online learning and offline self-learning effectively)	• Integrate both online learning and offline self-learning. • In the offline self-learning phase, students are given course-specific reading materials both before and after class with activities or assignments. • In the online teaching phase, educator should encourage discussions and group activities for students to exchange their understanding based on their reading. Thus, students will not learn ambiguous, fragmented, and surface knowledge. Instead, they will experience deep learning during the discussion. • Encourage global, community, and collaborative learning.

(Continued)

TABLE 1.1 *(Continued)*

Strategies	Description
7. Reflection (gauge students' understanding, learning outcome attainment and improvement for next session)	• Educator should provide feedback to students' assignments and know the learning cognitive levels of students. • Provide continuous feedback, quizzes, and assessments to ensure learning outcomes are achieved. • Allow students to provide suggestions and feedback on the learning session. • Reflect on ways to improve the next online learning session.

Source: Cheema (2020)

FIGURE 1.2 A Framework for Restarting and Reinventing School (Source: Darling-Hammond et al., 2020).

better place"(Wolff, 2020), (Djumalieva, J., & Sleeman, C., 2018). Surely there is no "one size fits all" solution, but aspects such as training, internet access infrastructures, hardware and software, digital literacy, and teaching and learning strategies for students and academics will be critical in this transformation. This chapter seeks to add to the analysis of the Covid-19 pandemic, envisioning it as an opportunity to foster the sustainable development of teaching in higher education. Thus, it intends to offer insights to be developed and furthered in future scientific works with distinct scopes (research paper, review, essay, etc.), aiming to understand this potential phenomenon, while concurrently providing contributions for policy implications, so that this process may be achieved effectively and efficiently. One thing I am sure of beyond any doubt is that machines perform much more efficiently as compared to human beings. But even still, it is practically impossible to replace humans with AIs, at least in the near future, because you can't build human intelligence into a machine, as it is a gift of nature. So, no matter how smart a machine can become, it can never replace a human. We might get terrified at the idea of being replaced by machines, but honestly, it is still a far-fetched notion. Machines are rational but don't have any emotions or moral values. They lack the ability to bond with human beings which is a critical attribute needed to manage a team of humans.

Yes, it is true that they can store a lot of data but the procedure of retrieving information from them is quite a cumbersome process, which is difficult compared to human intelligence.

The author of this chapter would like to conclude with Seneca's immortal words: "If one does not know to which port one is sailing, no wind is favorable".

REFERENCES

Aristovnik, A., Kerži˘c, D., Ravšelj, D., Tomaževi˘c, N., & Umek, L. (2020). *Impacts of the COVID-19 pandemic on life of higher education students: A global perspective.* Preprints.

Baker, T., & Smith, L. (2019). *Educ-AI-tion rebooted? Exploring the future of Artificial Intelligence in schools and colleges.* Nesta Foundation. Available at: https://media.nesta.org. uk/documents/Future_of_AI_and_education_v5_WEB. Pdf.

Buckingham, D. (2018). Defining digital literacy: What do young people need to know about digital media? *Nordic Journal of Digital Literacy, 2015*(4), 21–34.

Cheema, M. S. (2020). Covid-19 revolutionising higher education: An educator's viewpoint of the challenges, benefits and the way forward. *Life Sciences, Medicine, Biomedicine, 2020*(4), 1–6.

Darling-Hammond, L., Schachner, A., Edgerton, A. K., Badrinarayan, A., Cardichon, J., Cookson, P. W., Jr., Griffith, M., Klevan, S., Maier, A., & Martinez, M. (2020). *Restarting and reinventing school: Learning in the time of COVID and beyond.* Palo Alto, CA: Learning Policy Institute.

Djumalieva, J., & Sleeman, C. (2018). *Which digital skills do you really need? Exploring employer demand for digital skills and occupation growth prospects.* NESTA. Available at: www.nesta.org.uk/report/which-digital-skills-do-you-really-need/

Financial Times. (2017). *Education must transform to make people ready for AI.* Available at: www.ft.com/content/ab5daa64-d100-11e7-947e-f1ea5435bcc7).

Frey, C., & Osborne, M. (2017). The future of employment: How susceptible are jobs to computerisation? *Technological Forecasting & Social Change, 114*, 254–280.

Heilweil, R. (2020). *Why algorithms can be racist and sexist.* Vox.

Joshua, A. (2021). *Pickett-business.org.*

Kahn, K., & Winters, N. (2018). *AI programming by children.* Proceedings of the Constructionism 2018 Conference. Available at: https://ecraft2learn.github.io/ai/ (Accessed September 2021).

Karsenti, T. (2019). Artificial Intelligence in education: The urgent need to prepare teachers for tomorrow's schools. *Formation et Profession, 27*(1), 105–111.

Karsenti, T., Bugmann, J., & Gros, P.-P. (2017). Using humanoid robots to support students with autism spectrum disorder. *Formation et Profession, 25*(3), 134–136. http://dx.doi.org/10.18162/fp.2017.a135.

Kerry Rose. (2020). Available at: www.iotforall.com/the-place-for-artificial-intelligence-in-education.

Kuang, C. (2017, November 21). Can AI be taught to explain itself? *New York Times Magazine.* Available at: www.nytimes.com/2017/11/21/magazine/can-ai-be-taught-to-explain-itself. html (Accessed September 2021).

Lisa Plitnichenko. (2020). *E-learning industry.com.* Available at: www.jellyfish.tech.

Luckin, R., Holmes, W., Griffiths, M., & Forcier, L. B. (2016). *Intelligence unleashed. An argument for AI in education.* London: Pearson.

Merriam Webster. (2018). *Intelligence.* Available at: www.merriam-webster.com/dictionary/intelligenc (Accessed: 10 August 2021).

Parnas, D. L. (2017). The real risks of Artificial Intelligence. *Communications of the ACM, 10*(10), 27–31.

Popenici, S. A. D., & Kerr, S. (2017). Exploring the impact of Artificial Intelligence on teaching and learning in higher education. *Research and Practice in Technology Enhanced Learning, 12*(22), 1–13.

Rouhiainen, L. (2019). How AI and data could personalize higher education. *Harvard Business Review.*

Russell, S., & Norvig, P. (2010). *Artificial Intelligence: A modern approach* (3rd ed.). New York, NJ: Pearson Education, Inc.

Serpa, S. (2020). The global crisis brought about by SARS-CoV-2 and its impacts on education: An overview of the Portuguese panorama. *Science Insights Education Frontiers, 5,* 525–530.

Stone, P., Brooks, R., Brynjolfsson, E., Calo, R., Etzioni, O., et al. (2016). *Artificial Intelligence and life in 2030. One hundred year study on Artificial Intelligence: Report of the 2015–2016 study panel.* Stanford, CA: Stanford University Press. Available at: http://ai100.stanford.edu/2016-report.

Suresh Kumar. (2018). Available at: https://elearning.adobe.com/2018/10/role-artificial-intelligence-learning/.

Tesar, M. (2020). Towards a post-Covid-19 'new normality?' Physical and social distancing, the move to online and higher education. *Policy Futures Education, 18,* 556–559.

Unesco. (2019). *International conference on Artificial Intelligence and education, planning education in the AI Era: Lead the leap—Final report.* http://dx.doi.org/10.18162/fp.2018.a166.

Vinichenko, M. V., Narrainen, G. S., Melnichuk, A. V., & Chalid, P. (2021). The influence of Artificial Intelligence on human activities. In A. V. Bogoviz, A. E. Suglobov, A. N. Maloletko, O. V. Kaurova, & S. V. Lobova (Eds.), *Frontier information technology and systems research in cooperative economics. Studies in systems, decision and control* (Vol. 316). Cham: Springer. https://doi.org/10.1007/978-3-030-57831-2_60

Wolff, L.-A. (2020). Sustainability education in risks and crises: Lessons from Covid-19. *Sustainability, 12,* 5205. www.javapoint. (2021).

2 Overview of AI in Education

Archana Bhise, Ami Munshi, Anjana Rodrigues and Vidya Sawant

CONTENTS

DOI: 10.1201/9781003184157-2

2.1   AI IN EDUCATION: PRESENT AND FUTURE

2.1.1   Intelligence, Human Intelligence, and AI

While going through the literature, numerous definitions and meanings of the word intelligence are encountered, the most common being the "ability to acquire and apply knowledge" according to the Oxford dictionary. While according to the Merriam-Webster dictionary, intelligence is the "ability to learn or understand or deal with new situations or trying new situations". As shown in Figure 2.1, intelligence refers to a process where the power possessed by the brain helps to learn and understand situations, obtain knowledge based on learning, apply knowledge, and update knowledge depending on the experience gained after applying knowledge.

However, one important aspect of human intelligence is emotion, which is a vital characteristic directly affecting the performance of an individual while applying knowledge to a given situation. Human emotions can be impromptu based on external

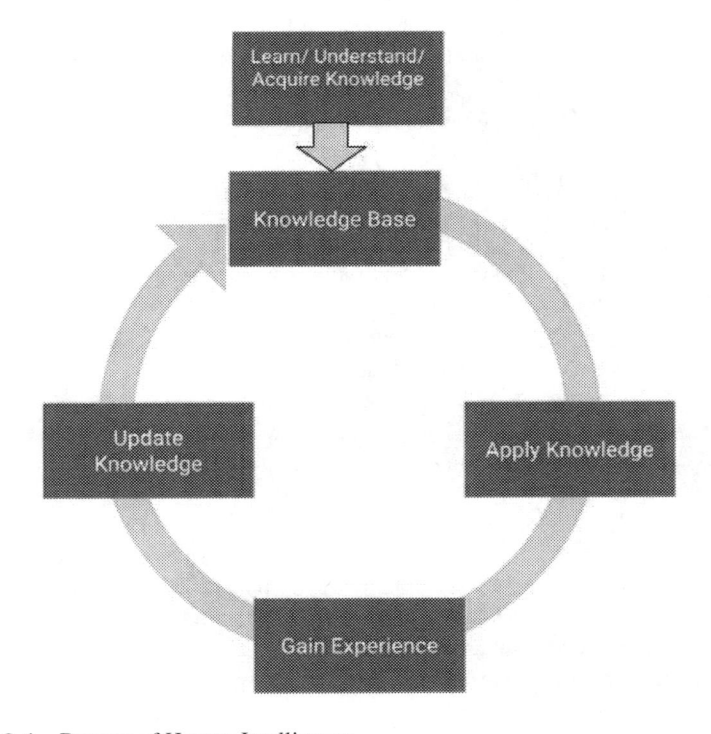

FIGURE 2.1   Process of Human Intelligence.

environmental stimuli or internal thought process (Darlington, 2018). Figure 2.2 depicts the role of emotions in human intelligence.

When an attempt is made to simulate human intelligence in machines such that machines start emulating human behaviour, it is called as AI. Such machines with the ability to execute tasks that typically requires a human are known as smart machines. Since AI systems are power-driven, they can perform assigned tasks efficiently and relentlessly without getting bored or tired like human beings. There are multiple definitions of AI available throughout the literature. Russell and Norvig (2010) categorize definitions of AI into four domains: thinking humanly, thinking rationally, acting humanly, and acting rationally.

AI acting humanly means performing actions like a human does. In 1950 (Turing, 1950), Alan Turing proposed the question of whether machines can think. He also created what is known as the Turing Test to ascertain whether a machine can act humanly. According to this test, if a machine is able to masquerade as a human by answering all the questions asked by the human interrogator, it passes the test. To act human, machines will require capabilities like:

• Natural Language Processing to communicate with the interrogator.
• Machine Vision to visualize its surroundings.

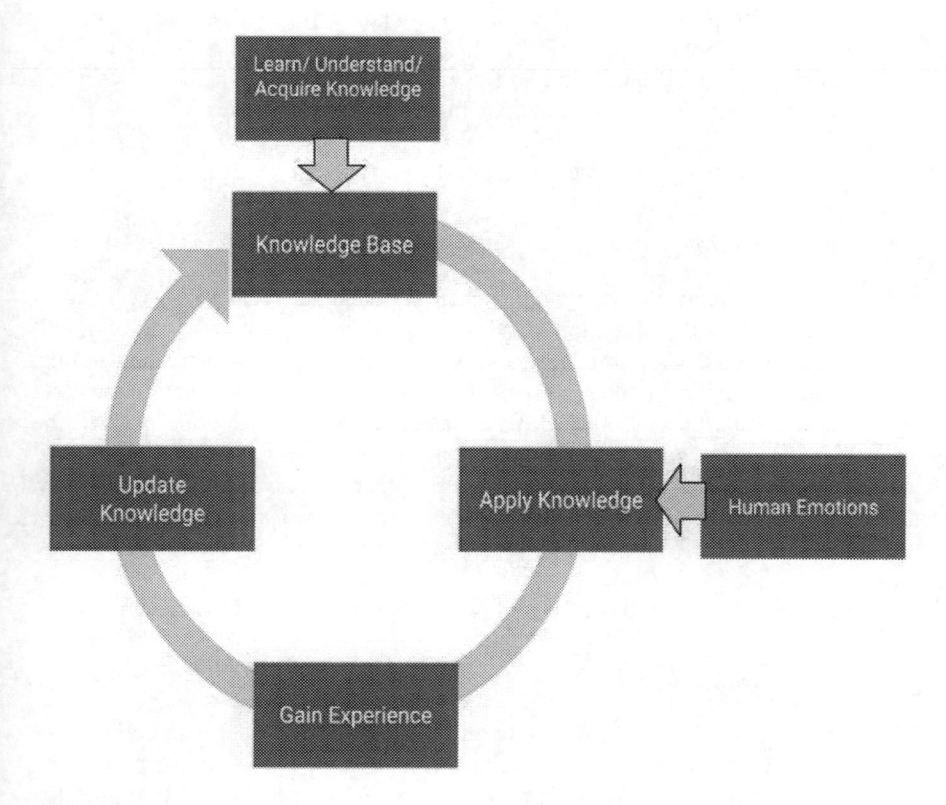

FIGURE 2.2   Process of Human Intelligence Influenced by Human Emotions.

- Knowledge base to store and retrieve information.
- Reasoning to respond to the questions of the interrogator based on information available in knowledge base.
- Machine Learning to adapt to the new circumstances, update the knowledge base and draw new conclusions.
- Motor Control or robotics to perform actions.

Thinking humanly, AI aims to mimic the human cognitive ability (Newman & Newman, 2020), to think abstractly, understand and solve complex problems, reason, and learn from experience. It gives the machine the ability to comprehend complex situations and find solutions.

Developing rational AI is based on a logical approach. Various kinds of logic theories including notations and rules of derivation were contributed by several Greek scholars like Aristotle (*History of logic—New World Encyclopedia*, no date). Rational thinking is an ideal way of thinking or right way of thinking—not always like humans. With the help of syllogisms, a kind of logical argument, a machine can be made to think rationally based on the structured arguments to yield an ideal conclusion.

Rational AI aims at building agents that will act rationally based on their external environment. Such rationally thinking agents must be capable of:

- Acting appropriately based on their surrounding perceptions, objectives, and knowledge base.
- Adapting to the changes in the surroundings.
- Learning from experiences and updating the knowledge base.

In this next section, various applications of AI will be discussed.

2.2 APPLICATIONS OF AI IN INDUSTRY

AI is being compared to electricity in its immense potential to transform every industry, every sector (Jewell, 2019). Slowly and steadily, AI has penetrated in various applications relevant in today's world due to its ability to solve repetitive, time-consuming, and sometime complex problems efficiently in comparatively lesser time without getting tired. While there are some apprehensions about adverse effects of AI, the presence of AI can be experienced in almost all the premises of business, like social media, finance, marketing, healthcare, entertainment, agriculture, cyber-security, military intelligence, robotics and automation, transport, tourism, gaming, astronomy, weather monitoring, e-commerce, education, and more. Some applications of AI are illustrated in Figure 2.3. AI has become an integral part of our daily lives to a great extent. In this section we will briefly go through applications of AI in a few industries.

2.2.1 AI IN MARKETING

Marketing is the business of promoting, advertising, selling, and conducting market research. All these tasks which were previously done by a team of people can now be easily replaced by a bot or an algorithm. AI is the catalyst changing the

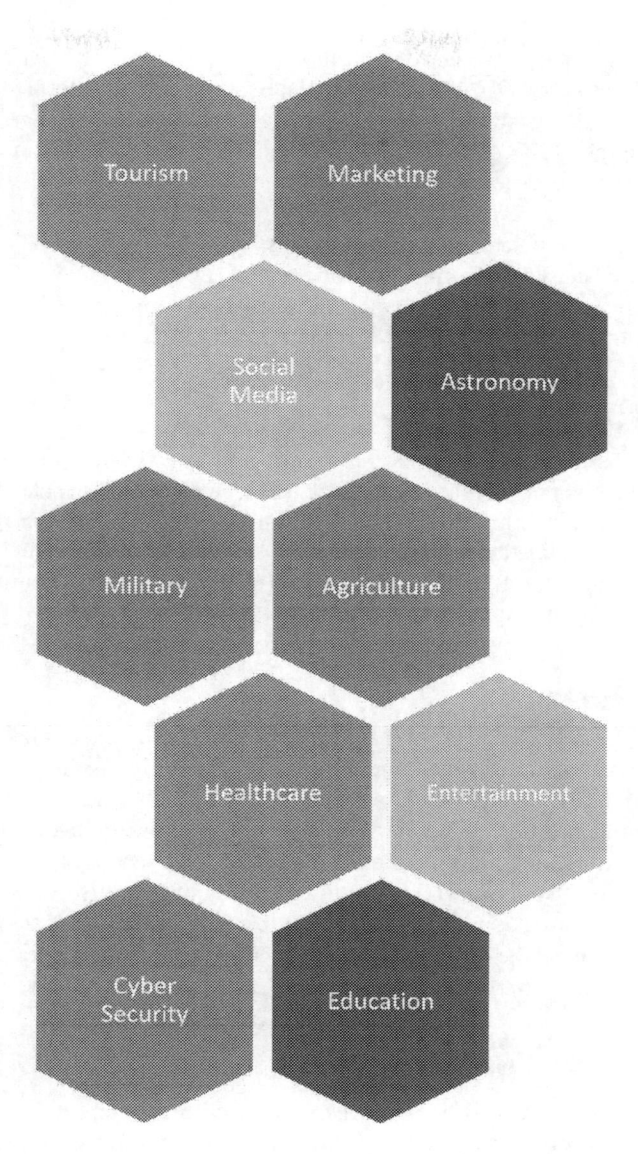

FIGURE 2.3 Applications of AI in Various Domains.

future of digital marketing. In our daily life, we come across examples of companies like Amazon, Flipkart, and Myntra which employ AI-based expert systems to promote relevant products based on the user's demographic like age, gender, ethnicity, income, previous purchases, and surfing history. Customer Relationship Management (CRM) software is needed for managing customer information and interaction. CRM software, propelled by AI capabilities, can become a robust data aggregation tool enabling businesses to collect, store, manage, and analyze data to improve customer interactions for a long-lasting relationship. Dominos (*Domino's ZERO Contact Delivery—Great Taste, Delivered Safe*, no date) has its own Chabot, Dom, which is able to remember the customer's previous orders, enabling the customers to order by simply messaging "PIZZA" through the Facebook messenger. In summary we can say that AI can be incorporated in every aspect of marketing lifecycles like:

- Identifying the potential customer and his/her needs.
- Reaching out to the identified potential customer.
- Converting the potential customer to an actual customer.
- Engaging the customer.

2.2.2   AI in Social Media

AI is one of the main mechanisms integrated in all the popular social media networks. Social networks like Facebook, Instagram, Twitter, and Snapchat have oodles of user profiles and their multimedia data. AI is needed to store and manage this huge data efficiently. Applications like Google Photos which are powered by AI make us nostalgic by creating collections of memorable photos and videos. Sentiment analysis is one of the promising premises of AI in social media.

2.2.3   AI in Healthcare

There are endless opportunities in the healthcare sector where AI is being used and can be used for leveraging systems to work more efficiently towards improving the overall experience. Explainable AI systems (Pawar, 2020) where the logic behind every predication is explored, aims to removing the complexity involved in the lives of patients, doctors, and administrative staff. Many healthcare organizations have employed AI-based systems for automating a variety of administrative applications such as appointment scheduling, billing, inventory management, and so on. Software powered by AI can aid the doctors in clinical decision-making to accurately diagnose ailments and provide appropriate consultation.

2.2.4   AI in Media and Entertainment

Like all other industries, AI has also revolutionized media and the entertainment industry. There has been a rise in the number of online platforms allowing users to upload online content created by them. While most of the time these contents are useful, sometimes inappropriate content is also available which might be harmful, insensitive, or violent, causing potential damage to its consumers like children. AI-based moderation technology and software can be useful to some extent to solve this problem. While accessing Over The Top (OTT) platforms like Netflix and Amazon, we have experienced personalized content recommendation based on parameters such as the consumer's liking, viewing pattern, culture, and age group. AI is also used for advertising, marketing, search optimization, promotional campaigns, classification of content, automated transcription, automated tagging, and more.

2.2.5 AI IN ASTRONOMY

Despite being a comparatively less-explored application of AI, the field of space exploration and astronomy has many exciting prospects. According to an article by Hausen & Robertson, (2020), a computer programme called Morpheus developed by University of California, Santa Cruz can be used to classify stars in the galaxy. Gravitational lenses enable astronomers to make observations about the dark matter distribution in galaxies. However, searching and tracking down gravitational lenses is a tedious process for astronomers, which can be easily done by AI based systems (Leary, 2017). New space missions can be planned and commissioned based on the analysis of previous studies done by AI-powered systems. Huge amounts of data and images generated by satellites can be analyzed using AI systems.

2.3 APPLICATION OF AI IN EDUCATION

The revolution that the education industry has seen in last decade has been influenced greatly by AI (Ayoub, 2020). AI has the untapped potential to bring about further transformation in every avenue of education and education systems like admissions, administration, teaching-learning methodology, evaluation, assessment, library management, training and placements, and so on. To follow are some areas where AI has a significant role to play in higher education.

- Personalized, customized learning experience based on student needs, caliber, grasp and understanding of material.
- Assistance for teachers for creating assignments, evaluation, assessment, analysis, task management and so on. Mechanize all administrative tasks so that the teacher can devote more time with students counseling wheres machine will not be of any help.
- Transcription in regional languages in presentations, videos and so on to facilitate better understanding to students.
- AI can help in automating the admission and enrollment process, thereby expediting the admission process, reducing workload on administration staff and decreasing human errors.
- Academic libraries of schools and colleges can be transformed into smart libraries by application of AI, thereby reducing the burden on the administrative staff and facilitating students to access a variety of material.
- AI can play a significant role in conducting and proctoring online examinations.
- Application of AI in school or college physical education may enrich the traditional physical education system.

2.3.1 SOME EXAMPLES OF AI-BASED TECHNOLOGIES USED IN EDUCATION

Jill, an artificially intelligent teaching assistant of Georgia Tech, is built on IBM's Watson platform. It is a bot that is trained to send introductory emails, reply to learners' queries about course syllabi, send reminders about assignment due dates, help students to solve design problems, and more (Korn, 2016; *Jill Watson, an AI Pioneer in Education, Turns 4 | School of Interactive Computing*, 2020; *Jill Watson: A Suite of Online Learning Tools—Design & Intelligence Lab*, no date).

Intelligent Tutoring Systems (ITS), which are designed to emulate human tutoring, provide customized learning and feedback experience to the students. Systems like MathiaU, ASSISTtments, Toppr, Thinkster Math, EdTech Foundary, are some examples of ITS fuelled by AI which aid the students in effective, personalized learning (Eddington, 2018; Walsh, 2019). With the help of ITS, teachers too can focus more on enhancing their teaching skills, refining their subject knowledge, and devoting time to research and development related activities.

Learning Management Systems like Blackboard, Google Classroom, and Moodle can act as a common interface among students, teachers and administration. LMS is a hub where teachers can upload their announcements, course outline, notes, presentations, course materials, assignments etc. Students are able to posts their solutions, answers, reports, projects, test, questions and so on in LMS. The administration can also communicate with students about fees, timetable, library, and other activities through LMS. Many schools and higher education institutions have developed their own LMS. For example, NMIMS University's SVKM Portal (Portal, 2021) caters to all of the abovementioned attributes of a LMS. IIT Bombay uses Moodle for its teaching-learning activities such as communicating with students, conducting assessments, uploading course material, organizing online discussion forums using Piazza which can be directly integrated into Moodle, and various other activities such as library services, uploading gradesheets, assignments, and so on (Moodle, 2021). Instructure Canvas is another very popular LMS used by many universities across the globe (Instructure, 2021). It provides facilities to form rubrics, create modules, develop calendars and schedules, conduct quizzes and other assessment, upload syllabi, perform data analytics, and more. Tools like Google Classroom, Microsoft Teams, Zoom, Adobe, and many other technologies can be integrated with Canvas API to create one centralized learning hub. Teachers can communicate with individual students, groups, or the entire class through messaging, audio notes, video, and more. Students can collaborate amongst themselves via chat group, video, and other messaging tools.

2.3.2   PROPOSED APPLICATIONS OF AI IN THE FUTURE

Although AI has become an integral part of education over the last two decades, there is still much more to explore. AI, when combined with Internet of Things (IoT) and data science, can make the education system richer by incorporating augmented intelligence for enhanced cognitive performance, student psychology, decision-making, and experiences.

- Wearable devices can be utilized for multiple applications. Smart watches can be used to automatically monitor the attendance of students without supervisor intervention. Wearable technology that incorporates sensors to monitor the heart rate and brain waves can be used along with AI to understand the physical and emotional state of students and teachers, for improving their interpersonal relationships, and for identifying states of depression, trauma and stress. Virtual reality headsets can be used to help students visualize concepts and processes, thus enabling faster understanding of subjects.

- Natural Language Processing (NLP) is a very useful tool for improving the language acquisition, learning and writing capabilities of students, particularly in e-learning. It can be used to increase the effectiveness of e-learning curricula, study material, presentation, and assessment in a large number of regional languages.
- Data analysis and AI can be combined to evaluate and proactively predict the performance of a student and identify major areas of strengths and weaknesses. Accordingly, AI-based suggestions of strategies for weak areas and opportunities to leverage strong areas can be provided.
- AI-powered face and voice recognition tools can be incorporated to judge the receptiveness and assimilation of topics being taught, and to provide real-time feedback to the teacher along with suggestions on the appropriate teaching methodology for a given topic.

2.4 ROLE OF AI FOR EDUCATORS IN HIGHER EDUCATION

2.4.1 ADMINISTRATIVE WORK

Some of the administrative activities that are carried out routinely by educators include question paper setting, assessment, grading, proctoring and plagiarism detection among others. These activities, though crucial for the stakeholders of the education system, are considered vacuous and mundane usually, and consume a considerable amount of time, energy and manpower to be properly executed. AI can be brought to the rescue here and these tasks can be executed with the help of intelligent computer applications, with only a supervisory role performed by the educator. The following paragraphs address the role of AI in some of these activities.

Question papers or their sets need to be set by the teacher many times during a semester for various assessments. Online examinations have taken preference over traditional methods in the modern world. This removes the need to print and produce physical copies of the question paper, transports it securely across campuses and does not limit the number of students appearing for an exam simultaneously. Subjective or multiple-choice questions (MCQ) can be set using platforms such as Google Forms, MS Forms, MS Teams, or Exam.net. As of now, there are very primitive, programmed AI-based features that assist in the setting of a question paper. The applications are capable of generating multiple sets of question papers from a question bank based on categories such as topic, sub-topic, difficulty level, marks allotted and so on. It definitely takes away a large burden from the shoulders of the teacher. It also ensures random question papers allotted to different students and reduces the possibility of answers being shared among students in an online examination. AI can further be improvised for incorporating advanced features that can gather information about the subject and help teachers by suggesting possible new questions. The AI-powered software can be made capable of generating a question bank on its own as per the given criteria, with the teacher only validating the question bank. A machine learning process can be used to train the software on the basis of students' performance to either increase or decrease the difficulty level of questions for the next examination.

Examinations are conducted to get a true measure of the knowledge assimilated by students. But this is possible only if exams are attempted in a fair and honest manner by the students. Hence, proper methods of proctoring need to be devised to deter students from sharing and copying answers. Remote proctoring methods enabled by AI use face and voice recognition systems to prevent impersonation. The candidate is allowed to appear for the exam only if he is correctly identified by the face recognition system. The same system can also continuously study the various facial movements and expressions of the candidate during the exam, and coupled with speech recognition techniques, malpractices adopted by the candidate can be captured. It is also possible to lock computer windows to just one screen so that copying from the same device can be prevented. Figure 2.4 depicts an AI-enabled remote proctoring system (*Eklavvya*, 2021).

Remote proctoring enhanced with AI, in spite of obvious and perceived advantages, comes with its share of concerns too (Nigam et al., 2021). Online proctoring systems, however intelligent, are not completely fool-proof. Another major issue is that of the security and privacy of students' personal data collected for verification. Hence, concerns over trust and ethical practices continue to plague existing remote proctoring software.

Assessment is another activity which requires hours of manpower that could otherwise be put to better use. At the same time, it is also one that demands discernment on the part of the assessor as it directly affects the grades of the students. In the

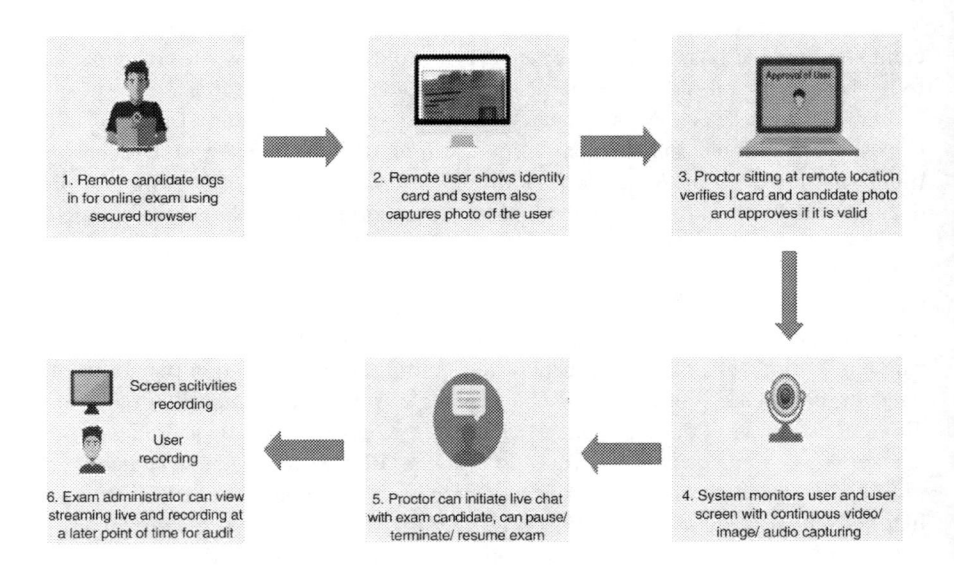

FIGURE 2.4   AI-Enabled Remote Proctoring System.

traditional method of assessment where the teacher manually reads the answers and makes a judgement on the quality of the answers, the process of awarding marks is dependent on various intellectual and emotional factors prevailing at that moment, such as the relative intelligence and performance of the class, the mental disposition of the evaluator, the effort that has gone into teaching the subject and hence the level of expectation of the evaluator, the amount of time that has been allotted to complete the assessment and so on. These influencers vary from time to time, person to person, and topic to topic. Hence, evaluation can be made more predictable, accurate and quick by transitioning to online evaluation using software trained for the purpose. AI based software such as Socrates, Google Forms, and Mentimeter are being used extensively for online question paper setting and assessment. For example, Mentimeter has built-in tools for evaluating a variety of skills like listening skills assessment, icebreakers, formative assessments, post-lecture surveys, and polls. It has been observed that most of the applications are more adept at evaluating MCQs and quizzes, and short answers to some extent. For subjective questions with long answers, online marking systems exist where the answer papers can be scanned and uploaded onto an application such as Mettl, and the online marking system allows the teacher to read and grade the answer sheets. The application is capable of doing all further tasks such as totalling marks and recording them in the exam portal. What all the online evaluation systems lack, however, is the versatility to think like a teacher and make decisions when it comes to evaluating elaborate, subjective answers. AI, NLP, and ML can bridge this gap by developing algorithms that can emulate the human thought process and perform assessments if certain criteria or key words are input to them. They can be made to have the capability to verify the quantity and relevance of the content, the grammar and the essence.

2.4.2 CONTENT DEVELOPMENT

Content development is as important as content delivery, if not more. AI is being used for the creation of "smart content" for all learners from kindergarten to the corporate. Smart content can range from digitized textbooks, guides that have been broken down content-wise into smaller chapters with the inclusion of self-assessment options, references to similar articles in other publications, customizable learning digital interfaces, content that can be made adaptive to the needs of the student, animated and visual representations of concepts, simulations, virtual laboratories, and more. For example, JustTheFacts101 (*CTI*, 2015) is a platform that allows the student to design his or her own curriculum, own customized lecture series, and view textbooks with a customized appearance and other features to make the learning process less time consuming and more appealing.

Other companies are creating smart digital content platforms, incorporated with content delivery, a large number of solved and unsolved exercises, real-time feedback and assessment options. Netex Learning (*Netex Learning*, 2021), for example, allows educators to design digital curricula and content across devices, integrating rich media like video and audio, as well as self- or online-instructor assessments, and a personalized learning cloud platform with apps, simulations, virtual courses, video conferencing, and other tools.

2.4.3   EFFECTIVE USE OF VIRTUAL PLATFORMS FOR TEACHING

The ongoing pandemic and the temporary closure of on-campus education across the world has given impetus to the use of virtual platforms for teaching. The education sector has woken up to the advantages and shortfalls of online teaching systems, and efforts are ongoing to improve the features every day. AI has played a pivotal role in enabling the teaching to continue without disruption though virtual teaching platforms. Some of the key benefits are:

- Sessions can be attended from anywhere using any device with internet connectivity.
- Sessions can be recorded and referred to for later use.
- No limit on the number of participants.
- Savings on infrastructure, electricity, water and other natural resources.
- Decreased crowd in transportation systems leading to reduced air and noise pollution.
- Savings in time and energy for both students and teachers.

Some of the disadvantages are:

- Hands-on training sessions in certain fields such as engineering, medicine and pilot training cannot be entirely replaced by virtual trainers or simulation softwares.
- Online training sessions for long hours can be taxing on the human body.
- Developing countries with insufficient telecommunications infrastructure and limited usage of electronic devices have difficulty in percolating the online teaching to remote locations.
- Peer review, motivation from contemporaries and inspiration from teachers experienced in face-to-face interactions is missing. The togetherness, joie de vivre and vitality of being with friends and classmates, and the ambience hence developed in the education system during curricular as well as extra-curricular activities is lost.

Online teaching platforms such as Kaltura, WebIQ, Zoom, and Blackboard Collaborate have been developed with the single purpose of catering to the education and training sector, and come replete with features such as video conferencing, collaboration and participation, online whiteboards, assessment tools, multimedia presentations, files upload and storage, chat, and so on.

2.5   ROLE OF AI FOR STUDENTS IN HIGHER EDUCATION

Learning is a process that develops a mind by gaining knowledge, skills, and expertise. Typically, the teaching and learning process occur in a classroom, with a trainer guiding and regulating the flow of information and knowledge to the students in a controlled environment in a predetermined time limit and at a specific location (Hasa, 2017). The learning process and its outcome is highly dependent on student behaviour, active participation in the learning process, and learning from their own mistakes. Moreover, a class consists of students of mixed ability, ranging from fast learners, slow learners, disengaged students, students who do not easily blend in with their classmates, and some having difficulty in understanding instructions or following textbooks. The "one size fits all" approach to education in a traditional classroom learning process thus faces challenges when applied to a range of students, thus limiting the learning outcomes. With technological advancements, virtual learning provides an alternate teaching-learning platform which has experienced consistent growth over the years and has emerged as a safe and viable option for continuity in education during Covid-19 pandemic lockdowns.

2.5.1 VIRTUAL LEARNING

Jethro et al. (2012) described virtual learning as a learning process that is created through interaction with network-based content delivered through digital platforms. Similarly, Engelbrecht (2005) described virtual learning as a learning environment that provides distance learning and teaching using electronic media such as the internet, CDs, mobile phones, or even television. On a virtual learning platform, teachers and students interact in real-time via chat, voice call, or video conferencing in a virtual class, which allows students to learn at any time of day from any geographic location, making learning more accessible and accommodating many students without difficulty. Providing personalized and timely feedback and guidance by the teacher has a positive impact on student performance, as it motivates students and makes learning processes easier. However, in a virtual learning classroom, the lack of face-to-face interaction restricts the scope of immediate personalized feedback.

AI mimics the cognitive functions of the human brain and adds value to the virtual learning process by personalizing education on an individual level. AI can search, find, calculate, strategize, and localize information scientifically and systematically much faster and more accurately than humans. Moreover, the AI engine analyzes learning patterns and provides recommendations. AI in virtual learning provides students with a more rewarding experience and freedom to learn at their own pace which leads to better learning outcomes. To follow are some of the ways AI tools may be applied to improve teaching-learning in a virtual learning environment.

2.5.2 DIGITAL COURSE CONTENT

Innovative digital course content and delivery plays an important role in virtual learning. The massive volume of educational information generated through online courses, teaching and learning activities brings value to the teaching-learning process. AI helps teachers to create, craft and customize the course material according to the learning capabilities and abilities of the students. Some of the benefits of digital course contents are shown in Figure 2.5.

AI empowers students with effective and interactive course contents thereby helping the learning process in following ways:

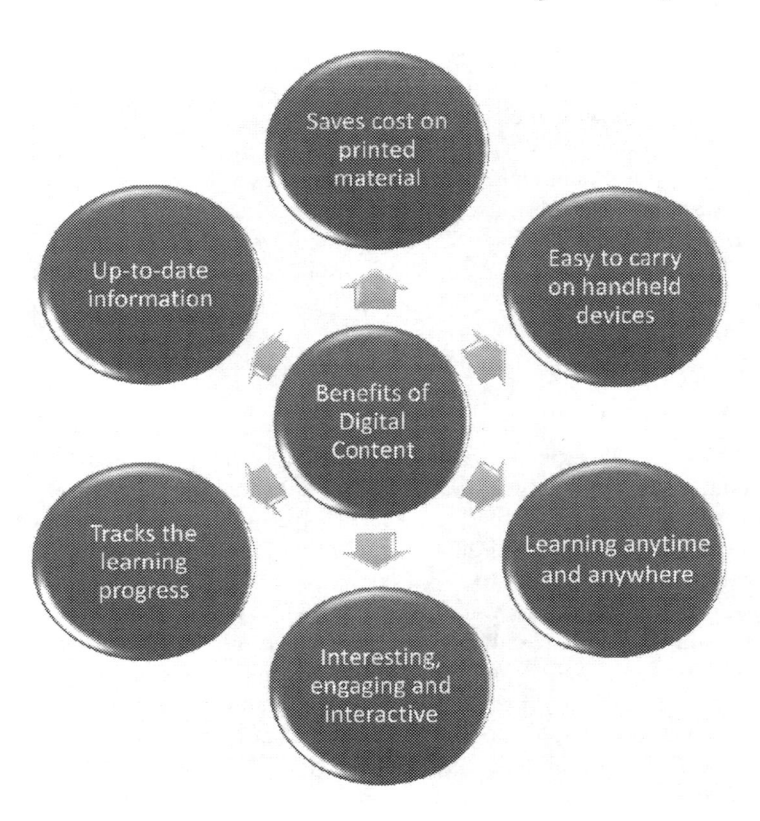

FIGURE 2.5 Benefits of Digital Course Contents.

2.5.2.1 AI-Powered E-textbooks

Reading is an essential component of the educational system. While the teacher uses textbooks for lesson plans, students read their textbooks to learn basic concepts and complete homework assignments that are reinforced in class by the teacher. This has been the mainstay of education for much of the past few hundred years. Textbooks have recently gone digital, and many classrooms are now supplementing or replacing traditional paper-based textbooks and educational contents with digital textbooks. Unfortunately, most digital textbooks are the scanned PDFs of the printed textbooks. Though digital textbooks are a more convenient alternative to their paper-based counterpart, they offer little benefit as a teaching tool. Moreover, a recent study discovered no difference in teaching effectiveness between paper-based textbooks and e-textbooks (Laketa & Drakuli, 2015). The study cited the lack of interactivity in e-textbooks, as well as their inability to personalize education to the needs of each student.

AI makes conventional educational material more intelligent, flexible and adaptive. Textbooks powered by AI and voice/natural language technologies increase student involvement in the learning process, educational effectiveness of the instruction and the ease of understanding by interacting with the e-textbook (Leddo John, 2020). AI powered e-texbooks provide some of the following fascinating features:

- Audio translations of the content, allowing students to listen rather than read.
- AI textbooks monitor each step of student learning and provide hints and feedback.
- As students learn a topic, they can verbally query the textbook to receive an answer to specific questions in the same way as using personal assistants on smart phones.
- The e-textbook assesses the students understanding on the topic by verbally asking questions, allowing students to answer verbally after the completion of the topic and immediately remediating the deficiencies.
- It adds a detailed explanation and practice questions for students struggling with any particular topic.
- Text summarization can reduce paragraphs of text into just the relevant bullet points for review.
- Adaptive textbooks also incorporate a learning method called retrieval practice that pops up occasional quizzes based on the material that students have already learned and enhances the student's ability to retain the concepts.
- Interactive visualization and animations.

Students often consider reading a passive activity and learn through graphics much faster than text. Visualization has the potential to present complex information concisely, while animation increases the effectiveness of visualization. Learning through visualization and animation enhances the active participation of the students and makes learning more enjoyable (Klerkx et al., 2014). Visualization enriches the teaching-learning process by presenting the abstract of concepts and their application in the form of graphs, maps, drawings, and images. Mendeleev's periodic table visualization is an excellent example of complex chemistry data that demonstrates the use of visualization for better understanding (Steele & Iliinsky, 2010). Visuals and animations can be used to demonstrate concepts not only in the classroom or in the laboratory, but they can also be included in online learning materials or made available as a reference resource to help students learn complex concepts.

Advent of modern softwares for interactive visualization enables the students to interact with the visuals, keeping them actively engaged in the learning process in a way which is not possible with static graphs. Interactivity provides the ability to identify, isolate, zoom, highlight, visualize information for extended periods of time and provide more valuable hidden insights. The number of visualizations available to educators has increased dramatically with the advent of tools such as flash animation and web applets. The interactive visualization has different levels of abstraction, complexity and interaction ranging from simple animations with the user having control of the playback buttons to sophisticated interactive visualizations in which the user can change the input data, environmental parameters, or constraints and see the results (Schweitzer & Brown, 2007). For example, GASP facilitates the visualization of geometric algorithms in a virtual classroom (Shneerson & Tal, 1997). The user need not have any knowledge of computer graphics to generate visualization. GASP provides every student with a control panel wherein the students are required to modify algorithm parameters without changing any code. Each student can see a distinct image of the same running animation in the lecture. The student can control, rewind and rerun the pace of the animation until the complex parts of the algorithm are understood completely. GASP employs a learning style in which the student is more constrained and must keep track of the teacher during the lecture. Algorithm Animation system, VisuAlgo, Explain Git with D3, Loupe for JavaScript, Alice 2, and jGRASP are a few examples of applications that provide animated visualizations for specific concepts (Majumder, 2019).

2.5.2.2   Video Captions

Adding captions to video lectures conveys the message to reach a broader range of students with difficulties understanding the language or with hearing disabilities. AI tools auto generate the subtitles for video lectures in the desired language thereby increasing student engagement and making the course content more accessible to students around the world.

2.5.3   AI-ENABLED LEARNING MANAGEMENT SYSTEMS

A Learning Management System is defined as an online learning technology for the creation, management and delivery of course material (Sabharwal et al., 2018). A LMS provides an interactive online environment shared by the teacher and the students enhancing and facilitating the teaching and learning. Through LMS, teachers can provide educational resources including syllabus, content resources and past exam papers. Students therefore do not have to worry about missing day-to-day notes and tutorials. Moreover, digital notes on LMS can be easily revisited, edited, updated and shared online. LMS encourages the communication between the teacher and students any time of the day, wherein the student can post the question at any time rather than waiting for an appointment with the teacher. The teacher also has the flexibility to reply at their own pace. Moreover, the inbuilt features of assessment and detailed tracking of student's progress not only allows the instructor to monitor and communicate the effectiveness of the writing process but also allows the learners see their progress in real time. Most universities are using an LMS such as Moodle (Dougiamas & Taylor, 2003), Sakai (Farmer & Dolphin, 2005) or WebCT (Goldberg, 1997), Blackboard (Blackboard Inc., 2006) to complement and support the teaching learning process.

Using AI algorithms with LMS automates the system, so it may perform without any human intervention improving the overall effectiveness of the training program. Han et al. (2017) discusses the use of AI in MOOCs for enhancing the existing MOOC experience through personalized learning, richer interaction opportunities and analytics of the resulting learning behaviour data. The ability of AI to develop intuition and prediction makes it ideal for virtual education via an LMS providing multiple benefits such as:

- A highly personalized training based on the cognitive, emotional, social, behavioural, and personality aspects of the individual. It builds its intelligence from the learning behaviour of each student and provides effective and targeted content to fit the learning style of each student.

- Grouping students based on their abilities and shortcomings through periodic and gradual evaluations of their performance and creating advanced and highly intuitive training programs.
- Optimizing learning materials based on the quality of previously taught material, student progress, and level of comprehension.
- Breaking down long-duration lectures and reading assignments into smaller sections based on the learning abilities of the student that can be easily understood by the student and suggests improvement areas through an automated constant assessment of their performance.
- LMS integrated with games, as a self-contained learning object, for increased student participation, engagement and rewarding the students for correct answers to keep them motivated as they progress through the course (Torrente, 2009).

2.5.4 CHATBOTS FOR VIRTUAL TUTORING

In digital learning, students can learn at any time of day and when doubts arise, the lack of a real teacher to answer the questions or the fear of asking the question in a real classroom should not prevent students from learning. It is highly important for the learner to get their questions answered immediately, as remaining unanswered they might later complicate the learning process.

A *chatbot* is software that simulates human-like conversations with users via text messages on chat. Chatbots such as ELIZA, ALICE, Alaude and Hex have been around for a long time (Smunty & Schreiberova, 2020). Conversational chatbots using AI have been an effective pedagogical agent in educational setting. AI-based chatbots, such as Google Duplex, AI-based Whatsapp chatbots, and so on offer personalized assessments, problem solving, and content recommendations for home learning and encourage them to complete tasks quickly. AI could also arrange for a live teacher to call the student and offer assistance, thus improving the efficiency of the teaching-learning process. AI-chatbots integrated as an interactive tutoring tool in digital classroom empower and simplify the learning process as follows:

- Provide the benefits of instant availability and the ability to respond naturally through a conversational interface, similar to an interview.
- Provide easy-going interactions with students so that they can be leveraged to support engagement, as well as setting out learning and engagement goals, strategies, and outcomes (Garrett, 2017).
- Provide a dedicated and unique learning environment for the student to respond to student's queries in the absence of an instructor.
- Answer basic and frequently asked questions from large pools of big data.
- Analyze data collected by the chatbot to assist the educator in improving the educational process and experience of the students.
- Arrange for a live teacher to call the student and offer assistance, thus improving the efficiency of the teaching-learning process.

2.5.5   E-Learning via AI Translators

Language often becomes a barrier in learning for international students and may complicate the learning process and affect the progress. Research shows that incorporating student's native languages increases the student's openness to learning by reducing the language barrier they encounter (Auerbach, 1993). However, it is extremely difficult to incorporate several different native languages in a traditional classroom especially when the teacher speaks only English.

AI translators are digital tools that use advanced AI to translate not only words that are written or spoken, but also the meaning and sometimes the sentiment of the message (Brenda, 2019). AI translators such as Google Translate translate text, handwritings and pictures in over 100 languages. Microsoft Translator provides live captioning for cross-language and even multilingual language understanding. Moreover, in-ear language translator devices not only translate conversations in real-time but also use AI to constantly improve translations. Integrating AI translators in digital classrooms for automatic text and language translation provides students with study material in their native language, thus bridging language and communication gaps.

2.5.6   Virtual Labs

Laboratory experiment are the most effective way to learn practical skills. Students can use laboratory equipment, analyze, interpret, and verify results in laboratories. However, open-source software makes almost all lab work available online and easily accessible to students. Students usually have concerns about using their laptops for performing laboratory experiments due to the inconsistent environments, incompatible hardware or issues with installing the required software on student machines due to corporate restrictions. Virtual Labs solves these problems by replicating the physical machines through a web app UI that allows students to use a wider range of hardware. Virtual Labs (VLAB), for example, is a multi-institutional open educational resource with over 1,650 virtual experiments mapped to the engineering curriculum (Raman et al., 2014). Experiments using VLAB have shown a positive impact on conceptual understanding, acquisition of conceptual knowledge, perceived enjoyment, motivation and the experience of using the labs in studies (Zacharia, 2007; Finkelstein et al., 2005; Josephsen & Kristensen, 2006). Moreover, virtual labs that use 3-D effects provide students with a better understanding of the lab setting for experimentation (Dalgarno et al., 2009). Some virtual labs adopt an inquiry-based approach that enables students to separate the variables that would be difficult in a physical lab setup (McElhaney et al., 2011). Systems like iSopt and Schatz's MOOC, for example, are heavily influenced by "inquiry-based" learning. Inquiry-based learning encourages students to work in a team; it challenges them with a question to which they formulate an answer (Waldrop, 2013). Furthermore, augmented reality (AR) and virtual reality (VR) can give virtual labs a more realistic feel. For example, a virtual and augmented reality (AR) environment for engineering education allows users to interact with 3D Web content (Web3D) using natural interaction techniques (Liarokapis et al., 2004). This not only enhances the laboratory experience; it can also be used to enrich traditional lectures by displaying multimedia content.

2.5.7 PERSONALIZED LEARNING

A teacher in a typical educational system teaches a class of 30 to 60 students, all along the same learning path. One of the main challenges experienced by teachers is to address the needs of the fast and slow learners in a class simultaneously. Personalized learning is an educational approach that customizes the learning plan of each student according to their strengths, needs, skills, and interests (Somasundaram et al., 2020). Personalized Learning requires someone to design a learning journey that is curated or created specifically for a particular learner and/or learning objective. However, creating a customized plan, tracking the progress and providing timely feedback to each student is a huge challenge.

An AI-based learning tool can process and reproduce information and knowledge a zillion times faster than an expert teacher. A well-designed, user-friendly AI-based learning tool can track the pace of learning of the students, their performance, identify the knowledge gaps, and assist students by customizing studies based on student's specific needs, preferences and learning styles. AI tools build a personalized study schedule and offer personalized content recommendations to each student thereby increasing their efficiency. An AI system can adapt, assimilate, and comprehend knowledge based on the results, edit history, and dynamics of a group of high IQ students thereby increasing the knowledge bank and creating a more powerful AI learning environment for brilliant minds. The use of AI provides insights to the teacher on the areas and subjects that require more attention, to optimize teaching and achieve desired learning outcomes. To follow are some of the AI tools that provide personalized help and guidance.

2.5.7.1 Course Recommendation System

Many universities are now using the Choice Based Credit System (CBCS) allowing students to select from a wide range of core and interdisciplinary subjects. The inter-domain learning platform enhances the skillsets of students, keeps them up to date with current developments, and improves employability. Selecting the appropriate subjects/streams is one of the key factors towards the career path of the student (Kapil Sethi, 2020). Inappropriate selection of subjects due to lack of information can lead to limited success in the selected stream. AI-based guidance systems designed to assist students to select a subject/stream based on information of student's interest, family background, previous education, successful scholars in the domain and other associated information can enhance career success.

2.5.7.2 Education for Students with Learning Disabilities

Students with learning disabilities often have difficulty in reading complex sentences and idioms found in text. In a typical classroom of around 60 students, educators may find it difficult to adapt the teaching style for each student, particularly students with learning disabilities. It can be difficult for the teacher to give the time, attention and instructions they require (Mathew, 2018).

AI has a lot of potential for helping students with disabilities learn more effectively. AI can help students with learning disabilities understand and engage with the material by replacing difficult texts with simpler, more understandable sentences.

Adaptive learning systems, multimedia learning systems, and games are also tools that enrich their learning experience.

2.5.7.3 Automated Grading with Personalized Feedback

Assessment and feedback are essential parts of any educational program. Assessments not only facilitate teachers in evaluating student's knowledge and skills, but they also help in shaping and directing the learning process. Feedback has a significant impact on student learning and has been described as "the most powerful single moderator that enhances achievement" (Hattie, 1999). Qualitative and quantitative feedback, obtained by comments and marks, guide students on the steps to improve and develop their capabilities. However, it is important to give timely feedback while the assessed work is still fresh in a student's mind, before the student moves on to subsequent tasks.

Students typically answer open-ended questions with a paper and pen in a traditional assessment conducted in an examination hall. A teacher's ability to provide timely and personalized feedback to each student is limited by such a traditional assessment, especially in a large class. The traditional assessment activity is demanding in terms of both time and effort and can be facilitated with AI-assessment technologies which will result in more timely and detailed feedback for students and the alleviation of teachers' workload. For example, a teacher can either take time to assess how each student is doing and address general learning barriers of the entire class or assess and provide feedback to each student individually. AI, on the other hand, enables students to receive personalized feedback at the same time.

AI has automated the assessment and grading process with multiple choice questions, fill-in-the-blank tests, and open-ended questions. Moreover, an AI-based assessment tool deploys natural language processing techniques, which checks the answers based on their meaning rather than comparing the words in the specimen. AI assessment tools provide a powerful impact on students learning by identifying the pedagogical materials and approaches adapted by the student, making predictions, providing timely interventions and optimizing the learning process (Lim et al., 2020).

2.5.7.4 Empowering Students with the 21st-Century Skill Sets

To succeed in school, work, and life in the 21st century, it is essential for students to excel not only in academics but also develop social-emotional and workforce skills. Beyond academics, AI tools can help students develop the attitudes, skills, and knowledge needed to understand and participate in a globalized world. Personalized learning AI tools allow the students to not only focus on content knowledge but also on the development of critical thinking through complex problems, innovative solutions, and the ability to collaborate and communicate across diverse teams.

2.6 CASE STUDY OF AI-BASED ASSESSMENT

2.6.1 Automatic Evaluation/Grading

Knowledge assessment is the most significant part of the education process, providing us with a measurable outcome about the knowledge gained by the student. It identifies learning gaps, helping in positive reinforcement to learners and encouraging them to perform better. With the advancement of technology, not only the ever-changing teaching and learning methods but the traditional pen and paper-based exams are also being replaced by automated online exams, making them more inclusive, accessible, and accurate. Moreover, automated online exams integrate the best practices of evaluation and examination developed over the years.

An online survey to understand students' perceptions of the context and attitudes towards online learning was conducted. The population for the study consisted of all students from four classes of undergraduate programme from Electronics and Telecommunication Engineering Department of the Mukesh Patel School of Technology, Management and Engineering (MPSTME), NMIMS (Deemed-to-be-University). Data was collected online through a questionnaire asking students to report the perceived benefits. The data was collected at the end of the second semester of the academic year 2020–21. The feedback comprised of four questions:

1. How would you rate the online exam? (5=very poor, 4=poor, 3=ok, 2= good, 1=excellent)
2. Which are the issues you had to deal with during the online examination?
3. What are advantages you associate with online exams (vs. traditional paper exams)?
4. Provide additional comments on online exams.

The answers to the open question were categorized and analyzed qualitatively. The study aimed to observe the attitude of students towards online exams. The respondents were requested to register their answers on a five-point scale varying from "very poor (5) to excellent (1)". Figure 2.6 shows the response to the first question that assessed the attitude of the students towards online exams in general. The respondents recorded mixed experiences of online exams with almost 52% of the students recording a strong positive, 32% an average, and 16% a poor attitude towards the online examination. This could be due to generally strong computer abilities of students across the globe, which allow them to use the online examination system with ease.

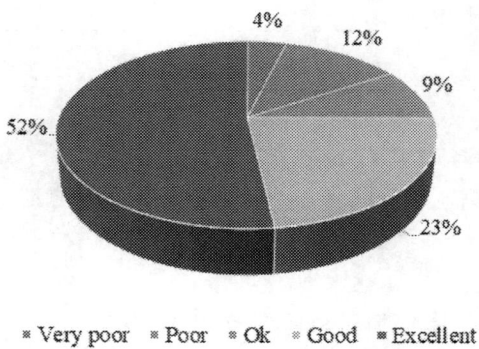

FIGURE 2.6   Students' Attitude Towards Online Exams.

The results to the question inquiring on the issues encountered by the students during the online exam were categorized as shown in Figure 2.7. Around 45% of the students reported technical glitches such as losing connectivity and limited bandwidth as major obstacles while taking the online examinations, while 57% of students reported typing speed, errors in typing, and the need to be computer savvy as obstacles in the online assessment process. Moreover, 31% of the students reported stress created by timed quizzes and the need to recall information in limited time.

The third question was focused on the students' perspectives about the benefits of online exams over traditional examination. The result to this question was categorized as displayed in Figure 2.8. Compared to traditional pen and paper assessments, nearly 81% of students reported rapid evaluation and display of results as the main benefit of automatic grading in an online examination system, while almost 56% of students reported it more interesting and convenient. Other benefits of online assessment mentioned by students included improved readability, and the ability to correct, modify, delete, and structure their test answers in a proper manner when compared

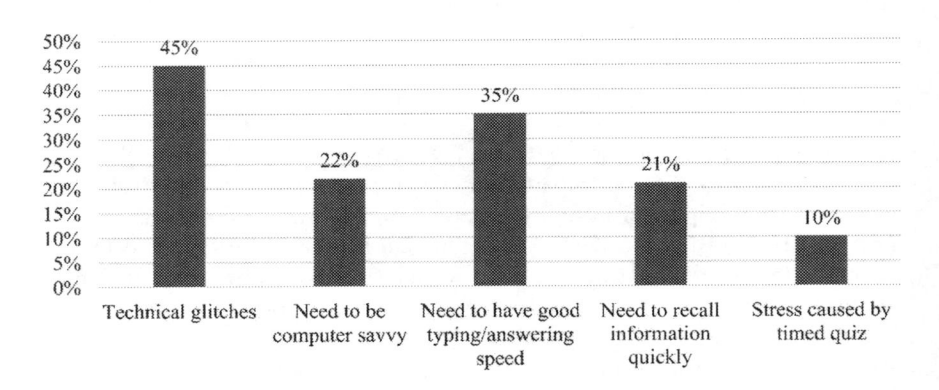

FIGURE 2.7   Difficulties faced by students during online exams.

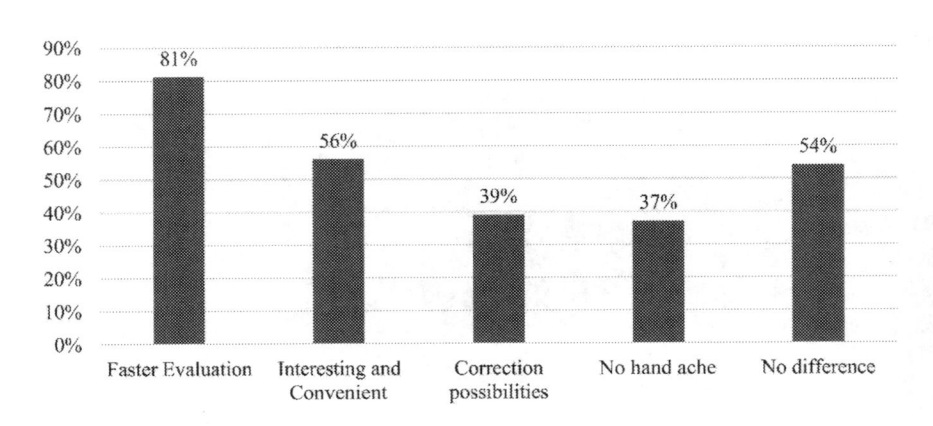

FIGURE 2.8   Benefits of Online Exams Reported by Students.

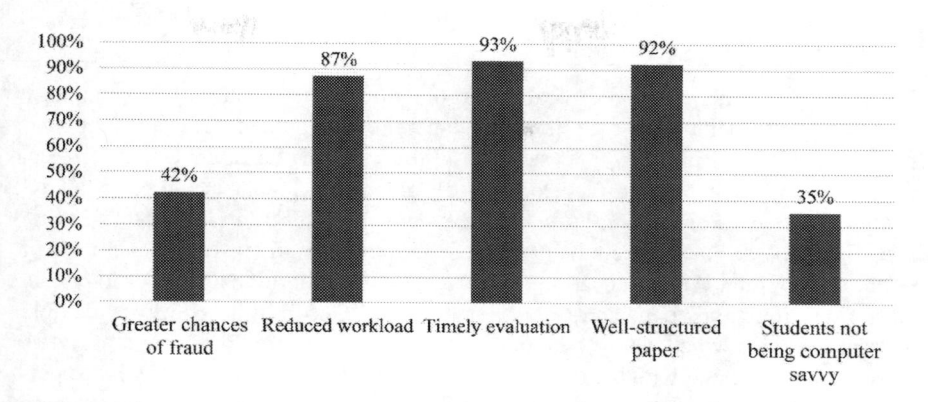

FIGURE 2.9 Perspectives of Teachers on Online Exams.

to paper-and-pencil exams. Few of them mentioned their hands were less tired than when completing a paper. Some students, however, found no difference between the online and traditional modes of exams. The responses were found to be consistent with the results of Frankl et al. (2012).

A similar survey was conducted to understand the perception of MPSTME teachers on online assessment, advantages of online assessment to traditional paper and pen examination and challenges faced by teachers while assessing students online. Overall, 32 MPSTME instructors replied to the questionnaire, distributed online using a Google form whose results are shown in Figure 2.9. From the results, it appears that the teachers have mixed opinions on the online assessment process. Around 87% of the teachers mentioned reduced workload and 93% reported timely evaluation as the major advantages of online assessment. Since the students take up the exam on their devices that are connected to the Internet, with all resources easily available to the students at their fingertips, around 42% of the teachers expressed concerns about fraud, and 35% of teachers expressed concerns about students not being computer and technology savvy. However, 92% of the teachers suggested that a well-structured paper with questions that are not easily found on the internet can prevent students from plagiarizing, compelling them to access in-depth-knowledge. Moreover, constructive and timely feedback aids students in identifying deficiencies and fostering learning to improve performance.

2.7 ETHICAL CONCERNS OF AI IN EDUCATION

AI is very useful in education. With the introduction of advanced technology in schools, there are questions raised regarding the social, economic and ethical impact the systems will bring. If not developed with good understanding, decisions about AI may have serious implications, especially where transparency and accountability are important. AI has enormous potential to threaten public safety, security and quality of life. Therefore, AI ethics is a very important part of AI applications in education (*Elaboration of a Recommendation on the ethics of Artificial Intelligence*, 2021). Additionally, AI is required to be designed ethically.

2.7.1   ETHICAL CONCERNS IN AI

Developing and maintaining ethical and professional AI standards for technology creators contributing to the design, development, deployment, management and maintenance of AI products and services is the need of the hour. UNESCO has, for the first time, decided to develop a legal, global document on the ethics of AI (*Elaboration of a Recommendation on the ethics of Artificial Intelligence*, 2021).

Companies such as Facebook, Amazon and Google are investing millions of dollars developing AI in education products. By 2024 the global market for AI in education products is expected to be worth around £4.5 billion. However, flexible learning systems which can be adaptable and personalized are not given importance by researchers. Along with the development of new AI technologies, guidelines, policies, and regulations should be framed to address the ethical issues which may come up with the use of AI in education.

Ethical concerns include fairness, inclusivity and diversity, an absence of bias, reliability, accountability, transparency and clear liability, clear privacy and data protection processes and security (Jameel et al., 2020). Broadly, the previous terms can be divided into two categories: data-centric and algorithm-centric. AI system developers and users are required to satisfy ethical requirements for positive impact of AI in education.

2.7.2   IS DATA ETHICAL?

Large amounts of data are collected for AI in education. We need to ask whether data is accurate, who has supplied data, who is the owner of data and how is data security maintained? Some of the ethical concerns related to data are:

- **Bias:** There is bias within data (that is, data that's skewed towards people and findings) as well as bias in the people who create the technology. Also, the data used to train AI systems can have biases (Pittinsky, 2020). Intelligent machines do not have a set of principles to identify bias like we humans do. If AIs develop a certain bias towards or against race, gender, religion or ethnicity, its use becomes questionable. For example, biased AI can prefer harmful outcomes by keeping female students, students in minority and other special students at risk. Therefore, those who work in AI research need to keep bias in mind when determining what data to use. Gender bias, racist and other forms of discriminatory assumptions are included in the data sets which are used to train AI systems. This is mainly due to intentional or unintentional approaches of the AI system's developer. AI trained with these datasets may generate biased predictions. A recent example is Amazon's hiring algorithm, which was criticized for being chauvinist, though there is no evidence that Amazon had any intention of being discriminatory. Without any explicit programming, the fact that more men had been successful in the past few years made Amazon create a model skewed towards replicating those results.
- **Data collection without consent:** This is directly related to data privacy such as issues of consent and privacy mainly due to data collection without first taking consent of the subjects regarding storage and sharing of personal

data. If this data is related to biometrics then it is legally considered sensitive data. It is expected that consent is taken from the subject before using samples to train AI systems.

- **Comprehensiveness of data:** There has been much discussion about this recently in terms of facial recognition. If an AI tool is developed using a fixed set of students from a geographical area then facial recognition may show accurate results based on the skin colour. AI tools developed by Google, IBM, Microsoft, and Face++ are much more accurate for light-skinned men than light-skinned women or darker-skinned men. Therefore, datasets used for training AI should contain samples from all quarters.
- **Security:** AI systems are vulnerable to cyber-crime which may cause damage if used maliciously. While stake-holders prefer dealing with an AI system that is faster and more capable than humans by orders of magnitude, cyber-criminals get more and more active with the intention of collapsing the systems. Globally, data security is considered to be a big challenge.

2.7.3 Is Algorithm and its Use Ethical?

With each year AI algorithms in education are getting more complex and accurate. However, the need of the hour is to use it for positive effects on learners. There are many ethical concerns related to AI algorithms in education (Zhou et al., 2020). Some of them are:

- **Opaqueness:** Generally, AI systems are not transparent and require minimal human intervention. For AI using deep neural networks, it can be impossible to find out why a machine is making the choices it makes. For example, in 2014 a computer proved a mathematical theorem. Though explanations of the software code were correct, it was beyond the capacity of humans to understand the same. Most of the time, a software developer is unable to comment on and justify the decision taken by AI system. Exploring the process of what is occurring within AI systems is not easy because developers use many variables which make them very complex. The whole point is to have computers do things that are not possible for human cognition. Trying to break that down ends up creating very crude explanations of what is happening and why.
- **Accountability:** Despite the good intentions of the people who develop and use AI systems in education, there are unintended consequences that are negative or that can even backfire. There is no way to find out who can be held accountable when decisions go wrong. Care should be taken to have AI developers become directly accountable to stakeholders and students of educational institutions. Technology creators who develop software tools might have used data which is focused on and drawn from a subset of the population that may not be useful for the students using AI. For example, AI learning systems that have been trained on students in urban area or students of high caliber may not have the same outcomes for students in rural areas. Or, AI systems developed for students a few years ago may not have the same efficacy for the students who are tech savvy and well-versed in digital platforms for learning.

- **Assessment without involvement of teachers:** Tools used for assessment collect information based on online interactions and capture key features teachers might notice in students. Consider the case where a student answers a question incorrectly. A machine will record a wrong answer. A teacher asks the same question in a different way and the student answers correctly.
- **Insufficient details of outcomes:** An AI system's outcome, justification of outcome and supporting details are completely dependent on developers' decisions. An AI system is expected to provide detailed information of students' learning ability, behaviour, background, and past academic record to the teacher which can be used for personalized teaching. Insufficient information about the outcome of AI systems may cause anxiety and displeasure between teacher and learner. For example, one predictive analytics tool estimated that 80% of the students in a Basic Electronics Engineering class would pass in this semester. This may not be news to professors. An AI system is expected to provide the justification of its predictions.
- **Prediction of career paths without involvement of human:** AI tools can take lots of data regarding career paths and make recommendations about what students can study. Conscious and unconscious prejudices of programme developers and those built into datasets train the software. For example, an AI might have been developed by a young person and thus decisions are based on his or her perception. Smart machines cannot outthink humans. They may end up making incorrect judgements.
- **Distractions:** It can be challenging for a teacher to monitor their students closely in a class so as to determine whether they are utilizing educational apps on their tablets or browsing social networking sites. Regular interaction with students and class participation could be possible solutions for such problems.
- **Plagiarism:** Due to the easy access of information and technological support provided, students blatantly copy information to hand in as a submission of their coursework. This not only breaches a student's integrity but also raises questions regarding the ethics of adopting AI systems within the education sector.
- **Legal concerns:** Different bodies of privacy and data-protection law should be considered before deploying any AI system. Data protection includes getting consent to disclose personally identifiable information. If institutions are sharing the information with outside parties and permitting them to access its information, it may be challenged if any information is incorrect.
- **Decision-making:** AI can automate decision-making which lowers the work burden of teachers and speeds up decision-making processes related to assessment of students. If an automatic decision is correct, society gets the benefit. However, if the decision is not correct, students' career may be affected. As AI developers are given more powers to make decisions, the need to have ethical standards becomes important. The ethical decision-making process might be as simple as following a programme to fairly distribute a benefit, wherein the decision is made by humans and executed by algorithms. If the decision is left to AI, it may entail much more detailed ethical analysis, even if we humans would prefer that it did not. This is because AI operates much faster than humans can and humans may lose control over

the execution of process before taking the decision. Some of the examples are cyberattacks and high-frequency trading. Situations such as these may worsen as AI expands its role in society.

2.7.4 UTILIZE AI WITH A GOOD CAUSE

There is no replacing the job that a teacher can do. AI can be used ethically in the education system. AI-based learning software assesses the extent to which it helps to promote or impede the satisfaction of a student's academic need. The key is to strike a balance between the support that technology can provide and the student's need. Presently, companies that develop or use AI systems largely self-police and rely on existing laws. AI should be tightly regulated. Policies to curb risks of unethical AI need to be in place. Software developers at companies consider their own liability from misuse before an AI tool is launched. It is difficult to anticipate and prevent every possible unintended consequence of their product. With the rapid rate of technological change, even the most informed regulatory body cannot keep pace.

Teachers can enable students to learn technology behind AI and ethical implications of new technologies so that they become responsible citizens and ensure that technology assists everyone for a better life. Several initiatives have been taken to promote ethical AI development in the field of business, education, politics, and sports. Globally there is a need to step up AI regulation to balance opportunities provided by AI systems and ethical concerns which are increasing day by day. AI standards will certainly minimize the risks of AI and ensure that AI technology is used to create a safe, secure, and fair environment for everyone. Many initiatives have been taken to address the ethical concerns of AI technology. One of the initiatives is to have AI ethics documents drafted by academic institutions. The professional organization IEEE is developing various technical standards for AI systems and corresponding ethical norms.

AI has vast potential, and its responsible implementation is up to us.

2.7.5 PLAGIARISM

Plagiarism is one of the most common issues in academia. Plagiarism is presenting someone else's work or ideas as your own, with or without their consent, by incorporating it into your work without full acknowledgement.

Evidence shows that the number of cases of plagiarism has increased exponentially in the last decade. Unfortunately, algorithms to evade plagiarism detection are also available. This has led to additional challenges for developers of plagiarism checkers.

Students copy text from other sources without citation. Students should be informed about the importance of writing using their own words and ideas. If a student is using someone else's work, he or she should include citations and reference the content of the author. Plagiarism is considered dishonest and unethical and is often punished by institutions (*Plagiarism—Wikipedia*, 2021).

In recent years, we have seen multifold development of AI technology in education. Unfortunately, plagiarism has also evolved with the same speed. Students can claim the content available on websites as their own. It is a matter of ethics. Teachers are required to create awareness among the students about academic integrity and creating one's own work—this has become a requirement in online learning.

For a teacher, the evaluation of assignments, project reports, papers, and tests is a routine and time consuming activity. The teacher is expected to apply plagiarism checks before finalizing the scores of students.

AI and machine learning algorithms have given new dimension to plagiarism detection. Machine learning and AI can identify plagiarism in texts and images within a very short time. These algorithms can also check whether a document is genuine or computer-generated.

Students can access free or paid services for writing project reports, formal communication, and assignments, and can get good score without any efforts. Various advanced plagiarism-checking detection platform like Copyleaks and Turnitin are available which can not only detect exact text matches and but also paraphrased content available on internet (Puiu, 2019).

Plagiarism detection applications use algorithms that search for identical or near-identical matches in the text between the database of key words/a sequence of words and contents to be checked. A similarity report can be reviewed before deciding the level of plagiarism or no plagiarism. Human intervention is required as a plagiarism checker may go wrong in declaring a portion of text as plagiarized or not plagiarized.

Unfortunately, the power of AI can be misused by students. Using OpenAI's GPT-2 algorithm, a student can generate a complete text with reasonably good quality. Even a learned reader would accept it as a genuine article. The flip side is that AI tools can mislead the users. A team at MIT used an AI algorithm to generate a meaningless essay using a set of commonly used keywords. The AI gave a high score to the meaningless essay.

Sometimes students resort to unintentional plagiarism. Instead of detecting plagiarism manually, an AI-based plagiarism test can be used to determine different forms of plagiarism. An AI plagiarism detection tool generates a plagiarism report which shows the sources from where the text is copied. The report shows percentages of similarity in comparison to self publications, external sources, and internet sources.

Amazon Web Services uses Neural Network and Natural Language Processing to provide plagiarism detection.

The AI-powered plagiarism checker helps teachers to scrutinize the work submitted by students. AI is the only solution to reduce, discourage and detect plagiarism in academia.

- **Turnitin:** This is an online plagiarism detection service that was launched in 1997 [8]. Many academic institutions are using Turnitin to check plagiarism of work submitted by students. It is based on machine learning, computer vision, and advanced AI and is used to find similarity of students' work and give real-time feedback. It has the ability of automatic grading to ease out the burden of teachers. A large number of schools and higher educational institutes are using Turnitin on a regular basis. It maintains a humongous database of documents, essays, research papers, academic articles and surveys, case studies, project reports, and theses.
- **Plagiarism of visual content**: In the digital era, plagiarism is extended to duplication of visual content as well. A person with wrong intention may

modify visual content and show it as his or her own. Students should be made aware that duplication in any form, whether text or images, falls under plagiarism. AI developers have developed online tools to find duplication of images over the web. The methodologies of duplicate image finders and plagiarism checkers are very similar.

- **Automatic Article Generator (AAG):** AAG is an online computer software which can access old and new academic articles, books and reports available on the web. The user inputs a research topic and/or a few keywords. The AAG algorithm searches these keywords and/or the research topic. It also replaces some of the words by synonyms before generating the final draft of the article on the required topic. An AAG-generated article has a good amount of information but it is not coherent. It may not convey the intended information. This is because AAG tools do not understand the meaning of words for a specific context. Therefore, they are not detected by plagiarism detection tools. Researchers are working on providing advance facility to AAG which will generate sensible and coherent articles. Sometimes, AAG-generated papers can be detected by analyzing the number of cited journals. AAG is based on an algorithm to search papers. It can cite a large number of papers unlike a genuine author who has a limited capability of citing the number of research papers. Guidelines are required to dissuade students from using AAG tools. A teacher is expected to make students aware of negative effects of the use of AAG. Therefore, briefing the students about ethical concerns of AAG and training them to identify computer-generated articles is very important.

2.7.6 How to Maintain AI Ethics in Education?

Students should not be given options to indulge in academic misconduct of AI ethics and plagiarism. Teachers can design assessment components in such a way that students do not get assistance from any AI tool. Assessment questions, projects topics and assignment work should be designed such that students are less likely to get the solution on the web or to outsource it to an outside agency.

In conclusion, it is very important to make students aware of ethical concerns of AI and plagiarism and its negative effects at the start of an academic term. They should be informed about the consequences they may have to face though their act which is unethical, even if the violation of academic misconduct is unintentional. Lastly, a training-the-trainers strategy is vital in this situation. Teachers must be given rigorous training on all the aspects of AI ethics, available AI tools, plagiarism, and plagiarism checkers to discourage students from indulging in academic misconduct.

REFERENCES

Auerbach, E. R. (1993). Reexamining English only in ESL classroom. *TESOL Quarterly, 27*, 9–32.

Ayoub, D. (2020). *Unleashing the power of AI for education | MIT technology review*. Available at: www.technologyreview.com/2020/03/04/905535/unleashing-the-power-of-ai-for-education/ (Accessed: 9 June 2021).

Blackboard Inc. (2006). *Blackboard building blocks*. Available at: www.blackboard.com/extend/b2/.

Brenda. (2019). *AI translators: The future of language learning?* Available at: https://oxford-housebcn.com/en/tag/resoures-learn-english/.

CTI. (2015). Available at: http://contenttechnologiesinc.com/ (Accessed: 23 June 2021).

Dalgarno, B., Bishop, A. G., Adlong, W., & Bedgood, D. R. (2009). Effectiveness of a virtual laboratory as a preparatory resource for distance education chemistry students. *Computers & Education, 53*(3), 853–865.

Darlington, K. (2018). *AI systems dealing with human emotions: How the future will be like with emotional machines.* Available at: https://www.bbvaopenmind.com/en/technology/artificial-intelligence/ai-systems-dealing-with-human-emotions/ (Accessed: 9 June 2021).

Domino's ZERO contact delivery—great taste, delivered safe. (no date). Available at: https://pizzaonline.dominos.co.in/? (Accessed: 9 June 2021).

Dougiamas, M., & Taylor, P. (2003). Moodle: Using learning communities to create an open source course management system. *EdMedia+ Innovate Learning*, 171–178.

Eddington, R. (2018). *Top 5 Artificial Intelligence tutors in education.* Available at: https://bigdata-madesimple.com/top-5-artificial-intelligence-tutors-in-education/ (Accessed: 10 June 2021).

Eklavvya. (2021). *Trends of Artificial Intelligence for online exams.* Available at https://online-examhelp.eklavvya.in/online-exams-artificial-intelligence-ai/ (Accessed: 30 June 2021).

Elaboration of a recommendation on the ethics of Artificial Intelligence. (2021). Available at: https://en.unesco.org/artificial-intelligence/ethics (Accessed: 23 June 2021).

Engelbrecht, E. (2005). Adapting to changing expectations: Post-graduate students experience of an e-learning tax program. *Computers & Education, 45*, 217–229.

Farmer, J., & Dolphin, I. (2005). *Sakai: ELearning and more.* Paper presented at the 11th European University Information Systems (EUNIS), Manchester, UK. R.

Finkelstein, N. D., Adams, W. K., Keller, C. J., Kohl, P. B., Perkins, K. K., Podolefsky, N. S., & Reid, S. (2005). When learning about the real world is better done virtually: A study of substituting computer simulations for laboratory equipment. *Physical Review Special Topics - Physics Education Research, 1*, pp. 010103:1–010103:8.

Frankl, G., & Schratt-Bitter, S. (2012). *Online exams: Practical implications and future directions.* Proceedings of the European Conference on e-Learning, pp. 158–164.

Garrett, M. (2017). Learning and educational applications of chatbot technologies. *Australia: Cinglevue* [Blog post]. Available at: http://www.cinglevue.com/learning-educational-applications-chatbot-technologies/

Goldberg, M. W. (1997). *An update on WebCT (world-wide-web course tools)—a Tool for the creation of sophisticated web-based learning environments.* Proceedings of NAU-Web'97 — Current Practices in Web-Based Course Development, Flagstaff, Arizona.

Han, Y., Miao, C., Leung, C., & White, T. J. (2017). Towards AI-powered personalization in MOOC learning. *NPJ Science of Learning, 2*(1), 1–5.

Hasa. (2017). *Difference between teaching and learning.* Available at: https://pediaa.com/difference-between-teaching-and-learning.

Hattie, J. (1999). *Influences on student learning.* Auckland: University of Auckland.

Hausen, R., & Robertson, B. E. (2020). Morpheus : A deep learning framework for the pixel-level analysis of astronomical image data. *The Astrophysical Journal Supplement Series, 248*(1), 20. doi: 10.3847/1538-4365/ab8868.

History of logic—New world encyclopedia. (no date). Available at: www.newworldencyclopedia.org/entry/History_of_logic (Accessed: 9 June 2021).

Instructure. (2021). Available at: www.instructure.com/en-au/canvas/resources/higher-education (Accessed: 20 September 2021).

Jameel, T., Ali, R., & Toheed, I. (2020). *Ethics of Artificial Intelligence: Research challenges and potential solutions.* 2020 3rd International Conference on Computing, Mathematics and Engineering Technologies: Idea to Innovation for Building the Knowledge Economy, iCoMET 2020. doi: 10.1109/iCoMET48670.2020.9073911.

Jethro, O. O., Grace, A. M., & Thomas, A. K. (2012). E-learning and its effects on teaching and learning in a global age. *International Journal of Academic Research in Business and Social Sciences*, 2(1), 203–210.

Jewell, C. (2019). *Artificial Intelligence: The new electricity*. Available at: www.wipo.int/wipo_magazine/en/2019/03/article_0001.html (Accessed: 9 June 2021).

Jill Watson, an AI pioneer in education, turns 4 | School of interactive computing. (2020). Available at: https://ic.gatech.edu/news/631545/jill-watson-ai-pioneer-education-turns-4 (Accessed: 9 June 2021).

Jill Watson: A suite of online learning tools—Design & intelligence lab. (no date). Available at: https://dilab.gatech.edu/a-suite-of-online-learning-tools/ (Accessed: 9 June 2021).

Josephsen, J., & Kristensen, A. K. (2006). Simulation of laboratory assignments to support students' learning of introductory inorganic chemistry. *Journal of Chemistry: Education Research and Practice*, 7(4), 266–279.

Klerkx, J., Verbert, K., & Duval, E. (2014). Enhancing learning with visualization techniques. In *Handbook of research on educational communications and technology* (pp. 791–807). New York: Springer.

Korn, M. (2016). *(99+) Original wall street journal story on Jill Watson | LinkedIn*. Available at: www.linkedin.com/pulse/wall-street-journal-story-jill-watson-ashok-goel/ (Accessed: 9 June 2021).

Laketa, S., & Drakuli, D. (2015). Quality of lessons in traditional and electronic textbooks. *Interdisciplinary Description of Complex Systems: INDECS*, 13(1), 117–127.

Leary, K. (2017). *Astronomers are using Artificial Intelligence to search for gravitational lenses*. Available at: https://futurism.com/astronomers-are-using-artificial-intelligence-to-search-for-gravitational-lenses (Accessed: 9 June 2021).

Leddo, J. (2020). Artificial Intelligence and voice powered electronic textbooks and electronic books. *International Journal of Social Science and Economic Research*, 5(1), 190–206.

Liarokapis, F., Mourkoussis, N., White, M., Darcy, J., Sidniotis, M., Petridis, P., Basu, A., & Lister, P. F. (2004). Web3D and augmented reality to support engineering education. *World Transactions on Engineering and Technology Education*, 3(1).

Lim, L.-A., Dawson, S., Gašević, D., et al. (2020). Students' perceptions of, and emotional responses to, personalised learning analytics-based feedback: An exploratory study of four courses. *Assessment & Evaluation in Higher Education*, 1–21.

Majumder. A. (2019). *A taxonomy for animation-aided visualization tools*, Master's thesis, Iowa State University.

Mathew, L. (2018). *Using Artificial Intelligence to help students with learning disabilities learn*. Available at: www.thetechedvocate.org/using-artificial-intelligence-help-students-learn ing-disabilities-learn/.

McElhaney, K. W., Linn, M. C., & Res, J. (2011). Investigations of a complex, realistic task: Intentional, unsystematic, exhaustive experimenters. *Journal of Research in Science and Teaching*, 48(7), 745–770.

Moodle. (2021). Available at: www.it.iitb.ac.in/moodle (Accessed: 20 September 2021).

Netex Learning. (2021). Available at: www.netexlearning.com/ (Accessed: 23 June 2021).

Newman, B., & Newman, P. (2020). *Cognitive ability—an overview | ScienceDirect topics*. Available at: www.sciencedirect.com/topics/psychology/cognitive-ability (Accessed: 9 June 2021).

Nigam, A., Pasricha, R., Singh, T., & Churi, P. (2021). A systematic review on AI-based proctoring systems: Past, Present and Future. *Education and Information Technologies*, 1–25.

Pawar, U., O'Shea, D., Rea, S., & O'Reilly, R. (2020). *Explainable AI in healthcare*. 2020 International Conference on Cyber Situational Awareness, Data Analytics and Assessment (CyberSA), pp. 1–2. doi: 10.1109/CyberSA49311.2020.9139655.

Pittinsky, T. (2020). *Algorithms and ethical diversity—IEEE technology and society*. Available at: https://technologyandsociety.org/algorithms-and-ethical-diversity/ (Accessed: 21 June 2021).

Plagiarism—Wikipedia. (2021). Available at: https://en.wikipedia.org/wiki/Plagiarism (Accessed: 23 June 2021).

Portal. (2021). Available at: https://portal.svkm.ac.in (Accessed: 20 September 2021).

Puiu, T. (2019). *AI: A tool both for detecting and enhancing student plagiarism.* Available at: www.zmescience.com/science/ai-a-tool-both-for-detecting-and-enhancing-student-plagiarism/ (Accessed: 21 June 2021).

Raman, R., Achuthan, K., Nedungadi, P., Diwakar, S., & Bose, R. (2014, November). The VLAB OER experience: Modeling potential-adopter student acceptance. *IEEE Transactions on Education, 57*(4), 235–241. doi: 10.1109/TE.2013.2294152.

Russell, S., & Norvig, P. (2010). *Artificial Intelligence: A modern approach. Third, prentice hall series in Artificial Intelligence.* Edited by S. Russell & P. Norvig (3rd ed.). NJ: Prentice Hall. doi: 10.1109/ICCAE.2010.5451578.

Sabharwal, R., Chugh, R., & Hossain, M. R., & Wells, M. (2018). Learning management systems in the workplace: A literature review. *International Conference on Teaching, Assessment, and Learning for Engineering*, 387–393.

Schweitzer, D., & Brown, W. (2007). *Interactive visualization for the active learning classroom.* Proceedings of the 38th SIGCSE Technical Symposium on Computer Science Education, pp. 208–212.

Sethi, K., Jaiswal, V., & MohdDilshad, A. (2020). Machine learning based support system for students to select stream (subject). *Recent Advances in Computer Science and Communications, 13*(3), 336–344.

Shneerson, M., & Tal, A. (1997). *Visualization of geometric algorithms in an electronic classroom.* Proceedings of the 8th conference on Visualization'97, IEEE Computer Society Press, pp. 455–458.

Smunty P., & Schreiberova, P. (2020). Chatbots for learning: A review of educational chatbots for the Facebook Messenger. In *Computers & education* (Vol. 151, p. 103862). Amsterdam: Elsevier.

Somasundaram, M., Mohamed Junaid, K. A., & Mangadu, S. (2020). Artificial Intelligence (AI) enabled intelligent quality management system (IQMS) for personalized learning path. *Procedia Computer Science, 172*, 438–442.

Steele, J., & Iliinsky, N. (2010). *Beautiful visualization: Looking at data through the eyes of experts.* Sebastapol, CA: O'Reilly Media, Inc.

Torrente, J., Moreno-Ger, P., et al. (2009). Integration and deployment of educational games in e-learning environments: The learning object model meets educational gaming. *Educational Technology & Society, 12*(4), 359–371.

Turing, A. (1950). *MIND, LIX*(236), 433–460. https://doi.org/10.1093/mind/LIX.236.433.

Waldrop, M. M. (2013). Education online: The virtual lab. *Nature News, 499*, 268–270.

Walsh, K. (2019). *Intelligent tutoring systems (a decades-old application of AI in education) | emerging education technologies.* Available at: www.emergingedtech.com/2019/12/intelligent-tutoring-systems-application-of-ai-in-education/ (Accessed: 10 June 2021).

Zacharia, Z. C. (2007). Comparing and combining real and virtual experimentation: An effort to enhance students' conceptual understanding of electric circuits. *Journal of Computer Assisted Learning, 23*(2), 120–132.

Zhou, J. et al. (2020). *A survey on ethical principles of AI and implementations.* 2020 IEEE Symposium Series on Computational Intelligence, SSCI 2020, pp. 3010–3017. doi: 10.1109/SSCI47803.2020.9308437.

3 Planning at the Edge of Tomorrow

A Structural Interpretation of Maltese AI-Related Policies and the Necessity for a Disruption in Education

Patrick Camilleri

CONTENTS

> Powerful technologies have always fascinated and frightened humanity, and humanity has gambled on them time and time again.
>
> (Metz, 2021, p. 10)

3.1 INTRODUCTION

Technology provisions what we as humans are physiologically missing. Technology has facilitated the process that empowered us from once merely competing with other species to essentially becoming the gatekeepers of the planet's resources for all. Technology has progressed in tandem with civilization effectively blending in within our mundane activities. In several instances it has also become the obvious if not consequential vehicle through which our lives are organized and ensue. Such technological determinism is also strongly disputed by discourse drawn from recursive dialogues taking place between users whose needs have direct influence on the design of the technology, and, the inherent structural qualities of technology that in several instances dictate the need for new requirements and skills (Mackenzie & Wajcman, 2005). It may also cause one to wonder if, as digital technology, and now, Artificial Intelligence (AI), are becoming more refined we are facing thinking machines that can seamlessly merge in, complement, mimic and unless properly catered for, determine what we can do.

Findings compiled by Benedikt and Osborne (2013) at Oxford University, Arntz et al. (OECD, 2016), and the McKinsey report on Automation and Employment (2017) all resonated the threat of automation on structured and manual jobs in general. Ironically the reports also came with a short-lived silver lining. In the same instances of predicting the vulnerable as the most vulnerable (Muro et al., 2019), the effect of automation on better educated workers was taken to be much less. However ongoing advancements and eventful technological disruptions brought about by AI and machines that can learn, reason and act for themselves (Muro et al., 2019) instilled concern for the permanence of seemingly unchallenged white-collar professions. Saying it differently, as technology such as AI progressively untethers itself further from human supervision, the issue is not anymore solely biased towards automation but more so towards autonomy (Gatt, 2021; Singh, 2020). Just, like in the case of the Japanese venture capital firm 'Deep Knowledge' that became the first company to name an Artificial Intelligence to its board of directors, AI is also provoking the reconsideration of what are exclusively human skills. Nonetheless, the issue here is not to shun away, because in context of technological progress we cannot. Rather, as the boundary on what pertains to human nature and what can be performed by the machine becomes more blurred, one wonders if ultimately humanity is after all really drifting towards a technologically deterministic future. Then again, the tendency to educate people according to what is deemed to be valuable at any particular time is a proof that historically, education has always risen to serve socioeconomic needs (Aoun, 2017). In context one may therefore ask:

> "What does it matter if students can't think, if machines will increasingly do the thinking for us?" But rather than raise the white flag on humanity and wipe the educational slate clean, we need to reconsider what we teach.

> (Aoun, 2017, p. 53)

3.2   CONTEXTUAL SETTINGS

Technological developments are undeniably outpacing educational provisions and policies to implement them according to perceived socioeconomic necessities (Facer, 2012; Rosenstock, 2014). There are several reasons, not the least being a misinterpretation of resistance from the teachers' side due to apprehension and lack of preparedness in embracing new pedagogies that relate better to our digitally mediated realities. This writeup therefore comes with the premise that the inclusion of digital technologies in formal educational settings is profoundly social, cultural and therefore a political concern (Selwyn, 2010).

Arguments are subsequently established around the following questions:

a. How are ICT policies reflecting socioeconomic foresights and plans in the wake of an AI permeated reality?[1]
b. Considering that education has socio political connotations, how are decisions dictated at the macroscopic level by policies, being associated with digital skills prerquisites in Initial Teacher Education (ITE) programs to effectively reflect digitally mediated socioeconomic requirements?

In this case, the theory of Structuration (Giddens, 2004) is primarily employed to interpret underlying political negotiations within an emerging AI-mediated reality. Structuration is then used to faciltate the insertion of the Structurational Model of Technology (Orlikowski, 2000) which, as exemplified through a recently published AI strategy, formalizes the need for an educational system directed to appease socioeconomic requirements in disruptive times. Incidentally this is happening in a country; which as decreed within the policy itself; is inclined towards a service-based economy. Malta is also investing substantial resources to embrace the benefits of Artificial Intelligence for our economy without neglect of its people as the best resource to harness this change in AI policy (Malta AI. Towards a National Strategy, 2019). Arguments then proceed to converge on the validity of how to best enhance the education process. When these are employed within the context of a digitally inspired society they also express how this evokes the need for a new pedagogy (Fullan & Langworthy, 2014) which has now to also accomodate for the perceived unsettling changes that AI will bring along.

The introduction of new technology generates multitudes of responses and interpretations (Griffith, 1999). Even if research has shown that technology-push and demand-pull can both be equivalent sources of successful innovation, inquiry into the absorptive capacity has mainly taken a technology-centric view (Schweisfurth & Raasch, 2018). As a consequence, digital technology seems to be endorsed as a natural component in the educational landscape (Selwyn, 2010). Then there is the tendency for the educational dimension of technology to be approached in a routine manner, rather than reflectively and selectively employed (Ibid., 2010). In this case whenever an ICT-related policy does not yield the expected outcomes, teachers are consistently put to blame (Gudmundsdottir & Hatlevik, 2018). In perspective, we discuss ITE programs (Tondeur et al., 2021; Liu, 2016) that support the setting of a paradigm for Education 4.0 (World Economic Forum, WEF, 2020a) in such Volatile, Uncertain, Complex and Ambiguous (VUCA) times (OECD, 2018) and expected transformations induced by technology (Caena & Redecker, 2019, p. 358).

3.3 THEORY: AN INTERPRETATION OF ICT APPROPRIATION

Technology is in search of applications. At the same time, societies are searching for solutions to problems based on intelligent information.

(Bangemann Report, 1994, p. 26)

---

1 It must be highlighted that whenever reference is made to the policy: Malta The Ultimate AI Launchpad (2019), it is addressed in the present. This does not only arise because of its very recent publication but also for the fact that what has been written is still unravelling and being implemented.

In the theory of structuration, Giddens (2004) explains how and why the dynamic interplay between macroscopic organizational features and human action at the microscopic levels make societies intrinsically temporary (Fay, 1997). Giddens (2004) also suggests that the: "the rules and resources drawn upon in the production and reproduction of social action are at the same time the means of system reproduction" (p. 19). Therefore the aspect of one recursively influences the features and outcomes in the other. Employing Giddens' (2004) discourse for technology yields a Structurational Model of Technology, specifically the Interpretive Flexibility of Technology (Orlikowski, 2000). In this case the process of adoption and adaptation to the technology in mundane life activities allows the users to experiment with and possibly even enact new qualities that go beyond the original deployment of the technology. Incidentally due to the context of use this may even invoke a redesign in the technology itself. Thus, the appropriation of ICT unequivocally drags all involved actors and tools in a sociotechnical process (Hussenot, 2008) where technological growth cannot be separated from the social context and where any entity involved can be a consequence for the shaping of the other (Warschauer, 2003; MacKenzie & Wajcman, 2005). As a matter of fact, segregating society from technology would not allow one to discern how the same structural qualities that define and create the indispensability of digital technology have their roots within socioeconomic needs (Castells, 2000).

3.4   THE CASE FOR MALTA: HARMONIZING EDUCATION FOR AI

For Malta, working towards nurturing an ICT committed culture is not a new endeavour. If necessity is the mother of invention then: "users are key producers of the technology by adapting it to their customs and values, ultimately transforming the technology itself" (Castells, 2002, p. 28). As experienced in the late 1980s, the formal recognition that isolation was not going to protect us from the effect of globalization instigated a thrust towards research and innovation. In 1988 the Malta Council of Science and Technology (MCST) was set up. It had the function of acting as an advisory board for the formulation and implementation of a national science and technology plan for modernization (Camilleri, 2017). Since then various implemented ICT-related strategies negotiated diverse aspects that reflected a shift from innovation to transformation (Ibid., 2017). As needs became necessities so did necessities justify outcomes for new opportunities. The closing of a chapter in the establishment of a Maltese ICT-oriented society was marked with a philosophical change of stance from, what to do 'with', to one of, what to do 'to' technology, moving the onus towards a more proactive and adventurous role where rather than limited to 'servicing with' ICT, Malta was envisioned as becoming a 'procurer of' ICT (Digital Malta, 2014). This stance became even more clear in the published high level policy document series: *Malta-Towards an AI Strategy* series[2].

---

2 It must be underlined that mirroring the fast pace with which technology and AI are rapidly unravelling in Malta; within a period of a year; three important AI related documents were officially issued. In 2018, the document referred as: Malta AI Vision document was launched as a precursor of the strategy that would have followed. The year 2019 saw the simultaneous launch of two documents:

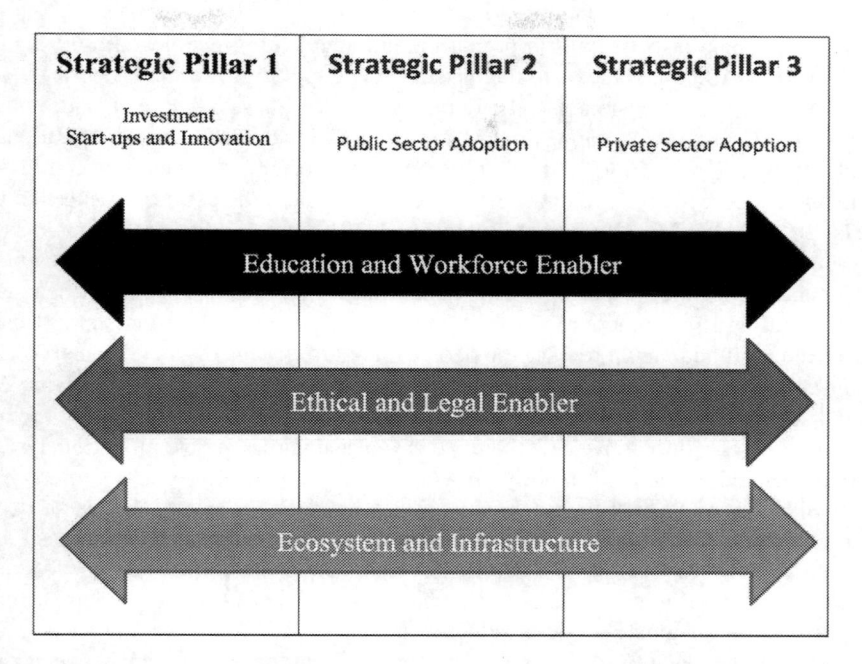

FIGURE 3.1 The Strategic Pillars and Enablers (adopted and adapted from: Malta the Ultimate AI Launchpad (2019).

In 2019, *Malta The Ultimate AI Launchpad* (Malta AI. Towards a National Strategy, 2019)[3] was launched. In the mentioned strategy it was clearly stated that in respect to AI, Malta rather than a follower should be a disrupter (p. 12). In this case disruption is a word that I believe can have various connotations. The Google dictionary (2021) defines disruption as a disturbance that can: '*interrupt an event or process*'. However it also associates the term with a "*radical change to an existing industry or market due to technological innovation*".

Undeniably the second meaning is more consistent with the portrayed strategy. Circumstantially and as detailed in Figure 3.1 to follow, the strategy is visualized to be attained through the investment of three strategic pillars and three transveral strategic enablers, one of which is education, directed to bolster the portrayed investments.

The strategy considers an AI-driven transformation to be disruptive. This is realized through the fact that as education permeats within each transversal enabler,

---

a. The strategy document coined as, *Malta The Ultimate AI Launchpad. A strategy and vision for Artificial Intelligence in Malta 2030*

b. A supporting framework referred to as: *Malta. Towards Trustworthy AI. Malta's Ethical AI Framework*

3 The Vision and Strategy for the inclusion of AI in Malta share names and pertain to: *Malta AI. Towards a National Strategy*. I will therefore distinguish the Strategy of 2019 from the Vision statement of 2018, by referring to it its title that is: *Malta the Ultimate AI Launchpad.*

education is also instrumental in preparing and augmenting peoples' attitudes and skills within different contexts to the onset of AI.

The strategy is envisioned to be deployed in two phases. These include what I interpret as firstly being a *harmonization stage* to include short term aspirational goals. If by 2022, the short term aspirational goals are achieved then they can instigate and support the second or *long term vision* of where the country would like to be by 2030 (Malta The Ultimate AI Launchpad, 2019). As pertains to the first phase, if it takes two to tango then any suggested implementation or political stance cannot be actualized without its people who are required to embrace and see the benefits of AI by putting it into practice. Thus, as ambitious plans require strong support it is also deemed fundamental to have the people on board. In this respect the short-term goals concerned with harmonization are directed to enhance awareness "of what AI is and some of the ways it can be used to benefit society" (Ibid., 2019, p. 16). Saying it differently, it is imperative to sell the idea by ensuring that people immediately see and understand the benefits underlying AI.

In his interpretation of the establishment of what defines society, Giddens (2004) proposes a dual nature or two facets of the same reality of society. Both structure; as defined by institutional properties; and subjective human action; denoted by the agency; are mutually liable to each others' influence. People are knowlegeable, can think, evaluate and are reflective (Ibid., 2004). Therefore any successful strategic implementations are also dependent on a mutual alignment between the envisioned or dictated structural properties and human action. When making reference to socio-technological relations, Orlikowski (2000) mentions the principle of sense-making. Users towards whom a technology is being directed to tend to firstly adopt it, but then there is meaningful employment. In this case the way that people eventually relate to, contextualize and give meaning to the named technology in their own realities will play an integral part in its appropriation and even giving reason for its valid presence or removal. Therefore this harmonization process is seen to be integral in the setting of the vision for AI adoption.

From an agency's point of view, the level and type of human interaction with technology involves the constitution of communication of meaning usually achieved through interpretations by all those involved. Therefore as part of the implementation process strategies are challenged, reaffirmed or changed accordingly by the human actors (Orlikowski, 1992).

Malta aspires to become the "Ultimate AI launchpad" . . . The Ambition is to create the conditions for AI to springboard from Malta to the world.

(Malta The Ultimate AI Launchpad, 2019, p. 15)

Ambitious plans also require strong support. One of the short-term goals is directed to enhance awareness, ". . . *of what AI is and some of the ways it can be used to benefit society*" (Ibid., 2019, p. 16). It must also be noted that in this age of the Fourth Industrial Revolution (4IR) nearly everyone will be working with AI (Gleason, 2018). Therefore, a critical mass of formally trained and licensed people that can constitute a workforce which can maintain an educated momentum where and when it is required is also very important. In the mentioned strategy the drive underlying the intensification towards communicating awareness and subsequent viability of AI employment is not being limited to educational institutions. Rather as a continuation of an ongoing, *"innovation-driven, service-based economy"* (Malta The Ultimate AI Launchpad, 2019, p. 12) and on the fact that: "The country *has now set out to establish itself in the field of AI"* (Ibid., 2019, p. 12), the need for education and training is seen to be pervasive in various socioeconomic contexts. Intrinsically the strategy is directed to:

a. Support and enhance all existing entities and its workers such as: Health, Financial Services, Gaming, Tourism, Real Estate, Advanced Manufacturing, Aviation and the Repair and Overhaul of Industries, and Educational Institutions.
b. Identify and support key sectors that can benefit from Research and Development and further support the drive of start-ups for development and Innovation.

While this strategy is particularly focused on embracing the impact of AI, it has not been written in isolation. Rather, as part of the implementation of the short-term goals, it underlines requirements of other previously released ICT strategy documents namely: *The Framework for the Educational Strategy for Malta 2014–2024* (MEDE, 2014) that emphasizes the importance of furnishing people with basic digital skills as part of their education. Consequently, the AI strategy looks to take their learning journey forward by augmenting their digitally mediated experiences with AI-infused personalized learning systems. Aoun (2017) refers to this principle of personalized learning as "humanics". Humanics, pertains to a form of procured complete or all-included personalised education. Humanics is therefore directed to equip future workers with the right knowledge prerequisites which, as Mitchell (2019) expresses will also include the skills and flair to employ what makes us human, differentiates us from machines, but in the meantime allowing us to work alongside machines.

For Aoun (2017) education is also its own reward. It equips people with the faculties of thinking. For many, it has also proven to be the key to climb the economic ladder. However, when the dictated socioeconomic requirements change, then so must we educate people in the *'subjects'* and *'disciplines'* that society deems valuable for its well-being. Incidentally the emphasis on 'subjects' and 'disciplines' arises because 'subjects' are suggestive of limited or defined bodies of knowledge. They essentially contribute towards the eventual formation of a discipline and therefore relate well within short-term timelines. On the other hand, a discipline is more in line and within the context of a vision. A discipline will therefore relate more to the formalization of an acquired belief or culture with a stronger sense of permanence. As pertains to the mentioned AI strategy, it is envisioned that the provision of AI modules of study within optional and compulsory higher educational courses will sustain the drive to endorse the 2030 vision.

Notwithstanding the expected extensive opportunities that AI through the right form of education can allow to be reaped, there are also extensive pitfalls for which education should prepare us. When comparing *The Future Jobs Report* of 2018 (WEF,

2018) with that of 2020 (WEF, 2020b), estimates become less optimistic. For the 2018 report (WEF, 2018) projections showed that automation would displace 75 million jobs but create 133 million more. On the other hand, the *Future Jobs Report of 2020* (WEF, 2020) projects that by 2025, automation and division of labour between humans and machines can displace 85 million jobs and generate 97 million others, a sheer 36 million less than previously projected. Undeniably this concurs with what Schwab and Davis (2018) stated that increased economic growth will no longer correspond with an increase in job growth. Incidentally for Malta this also brings about a change in focus between successive policies. Prior to the cited AI strategy, ICT policies were primarily directed to enhance the acquisition of digital literacy. They focused on adoption, accommodation and ultimately the enactment of constructive new qualities from digital technologies that would have facilitated productivity in step with developments in digital technologies and electronics. Nonetheless times have 'abruptly' changed again. We are now experiencing the onset of Industry 4.0. The term Industry 4.0 emerged as early as the 2000s from Germany's manufacturing industry (Gleason, 2018). Still, for the technology to materialize, we had to wait until the right capabilities for massive data collection and storing were developed, effectively triggering in machine learning. Now as we experience the fusion of several technologies, and where the digitization of everything is automating production and even automating the production of knowledge itself, how can education respond to the autonomy of our own machines?

The implementation measures as proposed by the aforementioned strategy are ramified and directed to meet widespread exigencies to "*create a sustainable and local engine for growth*" (Malta-AI-Vision, 2018), in diverse sectors of the Maltese society. This is deemed achievable where with education as its core driver, academia, startups and companies will contribute in an ongoing recursive dialogue of needs and procurement to sustain current and future generations of trained educated people.

> The Government recognises the critical function provided by the working population as well as the education system in Malta, which provides the fundamental function of sustaining current and future generations of trained and educated people.
>
> (Malta The Ultimate AI Launchpad, 2019, p. 37)

Notwithstanding all the groundwork being performed, the issue is that the 4IR is moving beyond the enhancement of the 3rd Industrial revolution (3IR). The 4IR is characterized with the unprecedented exponential advancement of new technologies that as Hussin (2018) considers are blurring the lines between the physical, digital and biological realities. Therefore the 4IR is unravelling itself in an unpredictable manner which is also not in line with previous industrial epochs (Schwab, 2018). Subsequently there can be potential gaps or misalignments in terms of forecasted economic requirements, and the projected and procured right-in-time skills. Therefore, what will the right form of education be like? As Kim (2017) pertinently asks in terms of higher education, "Why is it that books about technological induced economic change tend to focus on every other information industry except for higher education?"

3.5   EDUCATION 4.0: RIDING THE KONDRATIEFF WAVE

'Kondratieff waves' is a technical term employed by economists to describe the regular cycles of economic growth and recession. Implementation processes that involve technologies are always accompanied by processes of adaptation, accomodation and eventual enactment. Enactment is the last stage in the process after the introduction of a technology. It pertains to the underlying outcomes and the ways the named technology will be eventually employed in the required contexts. Exemplified contextual practices will either consolidate its implementation or request adjustment. Slotted in between enactment and implementation are processes of adaptation and accomodation. While theories of acceptance of technologies go beyond the remit of this article it still must be mentioned that adaptation and accomodation also involve the element of time. This can potentially provoke a misalignment between the expected innovation and productive growth as a result of the time required for training, eventual naturalization and enacted qualities that can enhance the constructive use of technology in question.

Reiterating what Aoun (2017) says, previous industrial revolutions relied on education to support development and benefit waves of innovation and transformation. But as illustrated in Figure 3.2 to follow, education is also an integral influencer and influenced component. It forms part of the recursive processes that exist between socioeconomic prerequisites that dictate how education and training should be shaped to sustain the same innovation that can only be sustained by the provison of education. Admittedly it seems that as it did in previous industrial revolutions, history is unconditionally repeating itself. On the other hand and through various aspects that include certification, teaching training research and innovation, education is being given a central enabling role. Recently published white papers by the WEF, such as:

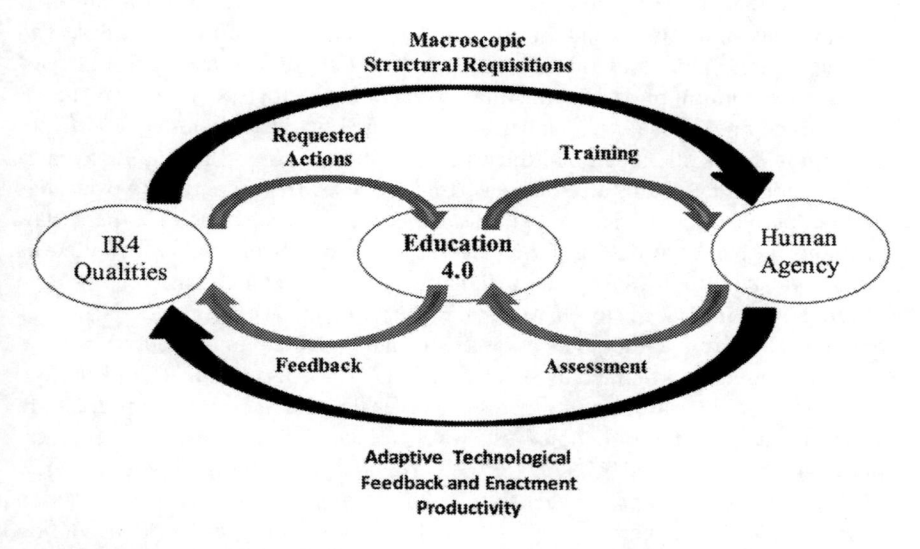

FIGURE 3.2   Processes of Recursivity and Ongoing Dialogues between IR4, E4.0 and Human Agency.

Accelerating Workforce Reskilling for the Fourth Industrial Revolution (WEF, 2017), tend to again provoke the redifinition of the recursive macro/micro cycles that are invariably associated with the 4IR. The quoted paper grants perspective to the dialogues enacted from the qualities of the 4IR. Specifically this necessitates an accelerated form of reskilling in the workforce that will shape gender and work in context of automation and now autonomous machines that define the 4IR. Therefore what stands in between this is a reshaped form of education coined as Education 4.0 or E4.0.

Hussin (2018) claims that Education 4.0 has become a shared buzzword between educationalists. The term E4.0 arises as a consequential reaction to the onset of the 4IR (Bonefield et al., 2020) and the effects that the latter has and will have on everyone's modus operandi. The definition for Education 4.0 is still *'nebulous'* (Bonefield et al., 2020, p. 224) and, as depicted in the previous figure, it is still emerging in response to a still unravelling 4IR. However, if as I believe, the 4IR is demarcated by the onset of machine autonomy then what form will Education 4.0 take? This arises from the fact that if the same autonomy in machines bestows them with decision-making elements, then they can encroach within professional blue-collar jobs. Thus: *"any predictable work-including many jobs considered 'knowledge economy' jobs-are now within the purview of machines"* (Aoun, 2017, p xii). This summons a transformative form of education where instead of preparing people for jobs that in the coming years will become obsolete or disappear under the rising tide of autonomous machines, specifically AI, people entering and/ or currently in the workforce have to be prepared for jobs that still have yet to enter the scene. Therefore Education 4.0 can be described as the outcome of the several consequential alignments that Higher Education must fulfill if it is to continue to prepare students for work (Aoun, 2017; Hussin, 2018; Bonefield et al., 2020). Katz and Associates (1999) considers that societal demands will shift practised education from conventional "just-in-case" forms, where knowledge is acquired before it is needed, to "just-in-time" in relation to when a person needs training. Yet Katz and Associates (1999) also goes further to depict "just-for-you" educational scenarios characterized with more personalized tailor-made forms that as Aoun (2017) considers, will be predisposed to be intrinsically defined by the person in need of specific training. Ironically albeit old, the arguments provoked by Katz and Associates (1999) are not outdated. It is also disconcerting because after more than two decades from their publication they still adjudicate themselves with so much meaning today. Undeniably just-in-case designs in education have been beneficial in bringing us to where we stand today. Perceived technology gaps between countries used to be resolved with an increase in tertiary education enrollments (Yeo & Lee, 2020) and as expressed in Figure 2 earlier, education accordingly adapted and procured workers to fit in prerequisites as defined by successive industrial eras (Aoun, 2017; Penprase, 2018). Paradoxically the same digital technologies that are underlining progress are also procuring uncertainty denoted as VUCA (Volatile, Unpredictable, Complex, Ambiguous) by Caena and Redecker (2019). Yeo and Lee (2020) employ the term, *'skill-biased technological change'* to refer to the non-neutral shift in production function which in the increasing presence of highly skilled workers is becoming the driving force underlying labor shifts and demands, as well as to the important relation between required knowledge acquisition and the nature of acquired skills

(Yeo & Lee, 2020). Consequently, beyond the fact that skill-biased technological change is the driver catalyzing the race between education and digital technology, it is also inciting the evolution of labor-skill composition (Ibid., 2020). Thus, as we contemplate on the underlying nature-nurture and demand-supply of future jobs and employability (Bonefield et al., 2020), it is natural to decide on how to design and deliver teaching and learning in sustainment of IE4 demands. The issue is therefore from where to start. This is a complex issue as it involves many stakeholders. Since policies confirm the importance of education as an enabler for transformation then teachers who are by nature at the core of formal education have the potential to be activators of meaningful learning (Caena & Redecker, 2019). Therefore, a good point to start from will be to consider Initial Teacher Education in context of seemingly adverse technological changes.

3.6 TOWARDS A NEW PEDAGOGY . . . PEDAGOGY 4.0

The employment of digital educational technologies (DET) is sustained by several stakeholders' interests. Selwyn (2013) considers DET as an arena where international and local organizational interests merge and ultimately stimulate their influence on the educational community comprised of schools, universities, teachers, academic researchers, and students. It is also known that DET implementation is liable to the actions from industry and commerce. In context, the extensive publication of policies that focus on the inclusion of ICT in education as a driver for the economy (Selwyn & Facer, 2013) is proof of the integral role that digital technologies are envisioned to play in the making and upkeep of contemporary societies. Notwithstanding that digital technology pervades in almost all aspects of our lives, speaking about DET today is still a complicated issue.

New technologies stimulate novel niches of opportunities. This also implicates the need for a well-defined strategic outlook on how best to proceed forward. However, the same technology-based changes are also provoking ambiguity in the meaning and the implications of terms employed particularly if: *"Changes in society and culture, based on new technology have effects on the terms used"* (Ilomäki et al., 2016, p. 656). Ilomäki et al. (2016), found discrepancies in the definitions of 'digital competencies', 'digital and media literacy', and 'eskills' (Ilomäki et al., 2016). This is reiterated by Oberländer et al. (2020) whom, albeit in terms of the place of work, still claim that there is still no agreed upon definition of the term Digital Competencies.

Arguably defining Digital Competencies in formal educational contexts and in the work-related environments may be different. However, I do not believe it should be so. Caena and Redecker (2019) refer to the importance of aligning teachers' profiles to 21st century competencies with living in the world, *"to navigate contemporary and future life, shaped by technology that changes workplaces and lifestyles"* (Ibid., 2019, p. 358). Likewise, in the Assessment and Teaching of 21st Century Skills (ATC215) framework, Binkley et al. (2012) outline the importance of an educational reform that sees a stronger association between what is being provided in formal educational contexts and the necessary skills for work. Subsequently, this places teachers and Initial and Continuing Teacher Education programs in Higher Education under scrutiny.

In 2008, Diana Laurillard expressed how education has been on brink of being transformed through learning technologies for decades. In terms of technology time-frames this may be taken to be outdated, however Caena and Redecker (2019) confirm that: *"The digital revolution has not yet been matched by mainstream transformations of education systems, teaching and learning in schools"* (p. 357). Despite the importance and, when compared to previous local policies, the urgency with which the inclusion of AI within formal educational contexts is being envisioned, reality shows that 'if' change happens within schools, rather than revolutionary it will be incremental (Selwyn & Facer, 2013). The effect of this misalignment in timing between technological developments and education is therefore making the preparation of future teachers an ongoing challenge (Tondeur et al., 2021). This can also be highlighted by the fact that they officially undertake and are responsible for the preparation of students for the future world of work. Thus, when teaching outcomes fail to meet expectations not only are teachers put to blame but substantial criticism falls on Initial Teacher Education (ITE) programs for failing to adequately prepare teachers for the complexity of the profession (Gudmundsdottir & Hatlevik, 2018). The issue here is not if students entering the profession should be prepared for an educational system that as time passes is becoming increasingly more digitally mediated (Starkey, 2020) but how best to do it. The dilemma is whether to prepare student teachers in ITE for the way schools are, or the way schools will be in the future (Gudmundsdottir & Hatlevlik, 2018); that is how (if we know how) the schools will change in response to technological innovation and change that AI is promising to induce.

On average, educational institutions take around 18 years to prepare people for careers that would span 40 years or more of their life. But as the future of many professions hangs in balance by the visible encroachment of AI and therefore autonomous machines, it seems that the education industry, including ITE, are now in a dilemma. Like Fullan and Langworthy (2014), Mitchell (2019) emphasizes that we cannot compete with machines. Ironically, the scientists themselves who have designed the original algorithms in the machines have been surpassed by the same machines. Mitchell (2019) also proposes that we, as humans, must enhance better what differentiates us from the machines. Autonomous machines execute precision; we as humans have unique flairs of creativity, flexibility and entrepreneurship (Aoun, 2017). ITE programs must therefore formalize Education 4.0 through what Fullan and Langworthy (2014) refer to as a new Pedagogy and which I call Pedagogy 4.0. Like teachers are at the core of the teaching and learning process, so should Pedagogy 4.0 motivate teacher preparation for the future. Literacy and numeracy are always important but ITE programs through Pedagogy 4.0 must be developed and focused on wider and maybe even less defined outcomes that lean towards deep learning qualities to include problem-solving, collaboration, creativity and thinking otherwise or differently from mainstream designs (Fullan & Langworthy, 2014).

Tondeur et al. (2021) state that when it comes to the employment of DET, a well-prepared teacher should be capable of understanding both the technology and the reality s/he will be using it in. The European Commission, Directorate-General for Education, Youth, Sport and Culture (2019) formalizes such qualities as interpersonal and meta cognitive skills namely focusing on the ability to learn-to-learn.

Saying it differently in today's reality we are spoilt for choice in an ocean of data that as previously expressed, gives purpose to the existence of machine learning and AI. It therefore transpires that teachers must be trained in decision-making qualities and be flexible enough to think outside the box to nurture qualities that enable them to reflect, decode and assimilate what is relevant for them in the huge data flow of information. They will therefore learn to appreciate what it means to become knowledge builders.

Admittedly, teacher preparation is a complex task (Tondeur et al., 2021). The attitudes that preservice teachers embrace with respect to DET is very much dependent on their personal experiences, efficacy and knowledge (Howard et al., 2019). To facilitate the development of digital competency within ITE, Tondeur et al. (2021) bring into focus the importance of personalized forms of instruction. When Aoun (2017) refers to HE and upskilling in the age of AI, he believes that one must first bring down artificial divides between knowledge and students, then break these chunks into smaller modular blocks. Finally, it must be seen that these blocks are accordingly assembled in line to targeted objectives. Incidentally this suggested personalized form of instruction echoes what Katz (1999) proposed when he referred to a "just-for-you" form of education which, he predicted, would have become more meaningful within mature digitally mediated environments. Again, this evokes the recursive dialogues (Giddens, 2004) where, working towards personalized attitudinal developments in DET within ITE programs at agency or microscopic levels (Tondeur et al., 2021) will align to required 21st century competencies (Caena & Redecker, 2019) at macroscopic levels.

3.7 CONCLUSION

This chapter has provided a snapshot of the rationale underlying the importance of investing in Higher Education to productively sustain economic change and take advantage of new opportunities only AI can bring. I dare say it is a snapshot because while education is taken to be instrumental to economic change, we are still poised at the edge of speculation on what machine autonomy will procure. However, the urgency for the requested transformation dictated within recent local and international policy documents (Caena & Redecker, 2019; Malta AI. Towards a National Strategy, 2019; OECD, 2018; UNESCO, 2019; WEF, 2018, 2020a) has also placed education systems, mainly teacher training under scrutiny (Gleason, 2018; Gudmundsdottir & Hatlevik, 2018; Starkey, 2020; Tondeur et al., 2021). In perspective of exemplifying Malta as a case study, the theory of Structuration (Giddens) has firstly served to interpret the required harmonization and recursive dialogues between policy presentation at institutional macroscopic levels and the agency response as represented by ITE programs and the teachers themselves within their personal microscopic realms. Secondly, in the process of highlighting the importance of aligning microscopic feedback to macroscopic requirements, the Maltese case served as a springboard to discuss the need for Higher Education institutions to accommodate for change on a global scale and within a new universal educational paradigm.

This new educational paradigm, referred to as Education 4.0, reflects the prerequisites of a rapidly evolving 4th Industrial Revolution and highlights the importance of shifting teacher preparation and student teaching from the provision of concepts and theories to an interoperability between digital skills, knowledge, and technologies. Keeping the prerequisites of the IR4.0 in focus, a newly required mindset referred to as Pedagogy 4.0 has also been discussed. The prevailing discussions on how to best prepare teachers to accommodate for new competencies and proposed innovative pedagogies were brought into focus. Arguments ultimately converged on how to best facilitate processes and empower teachers with the right attitudinal qualities and relevant skills to learn, unlearn and relearn. However, at the closure of this chapter, a pertinent question that I would have preferred not to conclude with but which I deem necessary is: *how can we hope to meaningfully relate and live if everything we are accustomed to is changing at an incredibly fast pace?* Like we have experienced before with the advent of the Internet, the actualization of enhanced connectivity between different devices and platforms has actively epitomized user-generated content causing great shifts besides other things in the implications of leisure, ethics, and socialization. Now we are being faced with AI. AI represents a different convergence and stimulus within a widening spectrum of frontier technologies (UNESCO, 2019). The outcomes of these technologies are unchangeably resonating with speculation both in terms of their validity and more so for their viability to sustain and satisfy socioeconomic needs and the sustenance of future welfare. Yet such uncertain indications also point towards defined needs within AI imbued jobs that instigate for unique human enabled qualities such as creative abilities (Ibid., 2019). Ultimately and admittedly paradoxically, the answer to the previously asked question is inherently through another question which is: *how to best improve learning outcomes within HE institutions and ITE programs and prepare the trainers for an AI imbued educational environment?* The need of the required creative abilities and therefore the procurement of such abilities can be achieved by transforming teachers' roles and preparing them for what it takes to work alongside AI-instilled educational environments. Yet in context of such uncertain times, I believe that one of the most, if not the most important competence that should be nurtured in preparation of our teachers, their students and eventual future workers for the oncoming uncertainties has been succinctly provided by Caena and Redecker (2019 as: *"the ability to adjust the skill performance to the demands of the situation".*

3.8   THEORETICAL AND PRACTICAL IMPLICATIONS

This chapter seeks to deliver an operational role to the theory of Structuration as presented by Giddens (2004). The theoretical lens of Structuration (Ibid., 2004) is inherently employed to facilitate the portrayal of an ongoing recursive dialogue between designated institutional requirements and users' contextual expected obligatory actions. However, the theory expresses how successful outcomes can only be achieved on the alignment between designated strategy policy requirements and complementing user actions. By specifically grounding the theory of Structuration (Ibid., 2004) in the context of an unraveling AI Maltese reality and augmenting the underlying theoretical framework with the Structurational Model of Technology as a modified theoretical lens (Orlikowski, 2000), a developing global digital technological reality is also brought into focus. Technological, and recently, digitally mediated, changes have always accompanied human development. Consequently the Theory of Structuration (Giddens, 2004) is instrumental in expressing macroeconomic inclinations and individual or group response to the implications, at the microscopic levels. Yet in this article the designated theoretical lens is also specifically being employed to define and give meaning to uniquely emerging digitally mediated traits that in this case are going beyond the accomodation to digital technology to acceptance and learning to enact new qualities to ride the wave of change that machine autonomy is bringing with it.

The recognition of AI and machine autonomy as an inherent threat to our current stability has also been notably recognized through the theoretical lens as a stimulus to change. The interpretive qualities within the Structurational Model of Technology (Orlikowski, 2000) have proven to be instrumental in understanding the intricate relation and interplay between the dictated criteria of the IR4.0, the reaction of the state as elucidated through policy analysis and the required spelled out qualities in Education 4.0 to mirror IR4.0 requirements. I however consider this article to be only a point of initiation and a point of entry into the human interpretation of an AI-inspired new era. I therefore see the employment of the Structurational Theory of Technology (Orlikowski, 1992, 2000) as being evocative if not provocative to further in-depth studies of human action, learning and also as a reaction to developing perceptions contextualized within an imminent AI-infused technological change.

REFERENCES

Aoun, J, E. (2017). *Robot-proof: Higher education in the age of Artificial Intelligence*. Cambridge, MA: MIT Press.

Arntz, M., Gregory, T., & Zierahn, U. (2016). The risk of automation for jobs in OECD countries: A comparative analysis. OECD Social, Employment and Migration Working Papers, No. 189. Paris: OECD Publishing. https://doi.org/10.1787/5jlz9h56dvq7-en.

Bangemann, M. (1994). *Europe and the global information society* (pp. 1–39). Recommendations to the European Council. Available at: http://aei.pitt.edu/1199/1/info_society_bangeman_report.pdf (Accessed: March 2021).

Benedikt, F. C., & Osborne, M. A. (2013). *The future of employment: How susceptible are jobs to computerisation?* Oxford Martin School Working Paper No. 7. Oxford: University of Oxford.

Bonefield, C. A., Salter, M., Longmuir, A., Benson, M., & Adachi, C. (2020). Transformation or evolution? Education 4.0, teaching and learning in the digital age. *Higher Education Practices*, 5(1), 223–246. https:// doi/full/10.1080/23752696.2020.1816847.

Binkley, M., Erstad, O., Hermna, J., Raizen, S., Ripley, M., Milleri-Ricci, M., & Rumble, M. (2012). Defining twenty-first century skills. In E. Care, P. Griffin, & M. Wilson (Eds.), *Assessment and teaching of 21st century* skills. *Research and applications*. Cham: Springer Nature. https://doi.org/10.1007/978-3-319-65368-6_1

Caena, F., & Redecker, C. (2019). Aligning teacher competence frameworks to 21st century challenges: The case for the European digital competence framework for educators (DIGCOMPEDU). *European Journal of Education, 2019*(54), 356–369. https://doi.org/10.1111/ejed.12345.

Camilleri, P. (2017). The ghost in the machine. A structurational interpretation of Maltese policies on ICT and education. *MRER. Maltese Review of Educational Research, 11*(1), 127–147.

Castells, M. (2000). *The rise of the network society* (2nd ed.). Oxford: Blackwell Publishing.

Castells, M. (2002). *The internet galaxy. Reflections on the internet, business, and society.* Oxford: Oxford University Press.

Digital Malta. (2014). *National digital strategy 2014–2020.* Aavailable at: https://economy.gov.mt/en/ministry/The-Parliamentary-Secretary/Documents/Digital%20Malta%20 2014%20-%202020%20(2).pdf (Accessed April 2021).

European Commission, Directorate-General for Education, Youth, Sport and Culture. (2019). *Key competences for lifelong learning.* Publications Office. Available at: https://data.europa.eu/doi/10.2766/291008

Facer, K. (2012). Taking the 21st century seriously: Young people, education and socio-technical futures. *Oxford Review of Education, 38*(1), 97–113. doi: 10.1080/03054985.2011.577951.

Fay, B. (1997). *Contemporary philosophy of social science. A multicultural approach.* Oxford: Blackwell Publishers.

Fullan, M., & Langworthy, M. (Eds.) (2014). *A rich seam. How new pedagogies find deep learning. Pearson.* Available at: www.michaelfullan.ca/wpcontent/uploads/2014/01/3897. Rich_Seam_web.pdf (Accessed March 2021).

Gatt, G. (2021). AI and the birth of true autonomy. Next 12. Reflections by some of Malta's thought-leaders on 2021. *Seedconsultancy.com.* Available at: www.gegegatt.com/ai-and-the-birth-of-true-autonomy/ (Accessed March 2021).

Giddens, A. (2004). *The constitution of society.* Cambridge: Polity Press.

Gleason, N. W. (2018). *Higher education in the Era of the fourth industrial revolution.* Palgrave Macmillan. doi.org/10.1007/978-981-13-0194-0.

Google Dictionary. (2021). Available at: www.google.com/search?dictcorpus=en-US&hl=en&forcedict=disruption&q=define%20disruption (Accessed April 2021).

Griffith, T. L. (1999). Technology features as triggers for sensemaking. *Academy Management Review, 24*(3), 472–488.

Gudmundsdottir, G. B., & Hatlevik, O. E. (2018). Newly qualified teachers' professional digital competence: Implications for teacher education. *European Journal of Teacher Education, 41*(2), 214–231. www.tandfonline.com/doi/full/10.1080/02619768.2017.1416085.

Howard, S., Tondeur, J., Ma, J., & Yang, J. (2019). *Seeing the wood for the trees: Insights into the complexity of developing pre-service teachers' digital competencies for future teaching.* ASCILITE 2019 — Conference Proceedings—36th International Conference of Innovation, Practice and Research in the Use of Educational Technologies in Tertiary Education: Personalised Learning. Diverse Goals. One Heart, pp. 441–446.

Hussenot, A. (2008). Between structuration and translation: An approach of ICT appropriation. *Journal of Organizational Change Management, 21*(3), 335–347. doi: 10.1108/09534810810874813.

Hussin, A. A. (2018). Education 4.0 made simple: Ideas for teaching. *International Journal of Education & Literacy Studies,* 82–99. http://dx.doi.org/10.7575/aiac.ijels.v.6n.3p.92.

Ilomäki, L., Paavola, S., Lakkala, M., & Kantosalo, A. (2016). Digital competence-an emergent boundary concept for policy and educational research. *Education and Information Technologies, 21,* 655–679. doi: 10.1007/s10639-014-9346-4.

Katz, R. N., & Associates. (1999). *Dancing with the devil. Information technology and the new competition in higher education.* A Publication of EDUCAUSE sponsored by Pricewater Coopers. San Francisco, CA: Jossey-Bass Publishers.

Kim, J. (2017). The fourth industrial revolution' and the future of higher ed. *Inside Higher Ed.* Available at: www.insidehighered.com/blogs/technology-and-learning/fourth-industrial-revolution-and-future-higher-ed (Accessed April 2021).

Laurillard, D. (2008). *Digital technologies and their role in achieving our ambitions for education.* London: Institute of Education.

Liu, S-H. (2016). Teacher education programs, field-based practicums, and psychological factors of the implementation of technology by pre-service teachers. *Australasian Journal of Educational Technology, 32*(3). https://doi.org/10.14742/ajet.2139.

MacKenzie, D., & Wajcman, J. (Eds.) (2005). *The social shaping of technology.* Berkshire: Open University Press and McGraw-Hill.

Malta AI. Towards a National Strategy. (2019). Malta the ultimate AI launchpad. A strategy and vision of Artificial Intelligence in malta 2030. In *Parliamentary secretariat for financial services, digital economy and innovation. Office of the Prime Minister.* Available at: https://malta.ai/wp-content/uploads/2019/11/Malta_The_Ultimate_AI_Launchpad_vFinal.pdf (Accessed March 2021).

Malta AI Vision. (2018). Malta AI. Towards a national strategy. In *Parliamentary secretariat for financial services, digital economy and innovation. Office of the Prime Minister.* Available at: https://malta.ai/wp-content/uploads/2018/10/Malta-AI-Vision.pdf (Accessed February 2021).

McKinsey & Company. (2017). *A future that works: Automation, employment, and productivity.* McKinsey Global Institute. Available at: https://www.mckinsey.com/~/media/McKinsey/Featured%20Insights/Digital%20Disruption/Harnessing%20automation%20for%20a%20future%20that%20works/MGI-A-future-that-works_Full-report.ashx (Accessed March 2021).

Metz, C. (2021). *Genius makers. The Mavericks who brought AI to Google, Facebook, and the World.* Dutton: Penguin Random House.

Ministry for Education and Employment; MEDE. (2014). *Framework for the education strategy for Malta 2014–2024: Sustaining foundations, creating alternatives, increasing employability.* Available at: https://education.gov.mt/en/resources/Documents/Policy%20Documents%202014/BOOKLET%20ESM%202014-2024%20ENG%2019-02.pdf (Accessed April 2021).

Mitchell, M. (2019). *Artificial Intelligence: A guide for thinking humans.* New York: Farrar, Straus and Giroux, Broadway.

Muro, M., Whiton, J., & Maxim, R. (2019). *What jobs are affected by ai? Better-paid, better-educated workers face the most exposure.* Metropolitan Policy Programme at Brookings. Available at: www.brookings.edu/research/what-jobs-are-affected-by-ai-better-paid-better-educated-workers-face-the-most-exposure/ (Accessed March 2021).

Oberländer, M., Beinicke, A., & Bipp, Y. (2020). Digital competencies: A review of the literature and applications in the workplace. *Computers & Education, 146*(2020), 103752. ISSN 0360–1315. https://doi.org/10.1016/j.compedu.2019.103752.

OECD. (2018). *Future of education and skills. Education 2030. The future we want.* Paris, France: OECD Publications. Available at: www.oecd.org/education/2030/E2030%20Position%20Paper%20.pdf (Accessed March 2021).

Orlikowski, W. J. (1992, August). The duality of technology: Rethinking the concept of technology in organizations. *Organization Science, 3,* 398–427. The Institute of Management Science USA.

Orlikowski, W. J. (2000, July–August). Using technology and constituting structures: A practice lens for studying technology in organizations. *Organization Science, 11*(4), 404–428. Published by INFORMS.

Penprase, B. E. (2018). The fourth industrial revolution. In N. W. Gleason (Ed.), *Higher education in the era of the fourth industrial revolution.* Palgrave Macmillan. doi.org/10.1007/978-981-13-0194-0.

Rosenstock, L. (2014). Fundamental change in education. In *A rich seam. How new pedagogies find deep learning.* Available at: www.michaelfullan.ca/wpcontent/uploads/2014/01/3897. Rich_Seam_web.pdf (Accessed April 2021).

Schwab, K., & Davis, N. (2018). *Shaping the future of the fourth industrial revolution.* World Economic Forum. Crown Publishing. Group, a division of Penguin doi: 10.5555/3312394.

Schweisfurth, T. G., & Raasch, C. (2018). Absorptive capacity for need knowledge: Antecedents and effects for employee innovativeness. *Research Policy, 47,* 687–699.

Selwyn, N. (2010). Looking beyond learning: Notes towards the critical study of educational technology. *Journal of Computer Assisted Learning*, *26*, 65–73. https://doi.org/10.1111/j.1365-2729.2009.00338.x.

Selwyn, N. (2013). *Education in a digital world. Global perspectives on technology in education*. London: Routledge.

Selwyn, N., & Facer, K. (2013). The need for a politics of education and technology. In N. Selwyn & K. Facer (Eds.), *The politics of education and technology. Conflicts, controversies and connections*. New York: Palgrave Macmillan.

Singh, S. K. (2020). The generation of Artificial Intelligence and the beginning of true autonomy. *East Facts*. Available at: www.eastfacts.in/2020/04/AI-Evolution.htm (Accessed March 2021).

Starkey, L. (2020). A review of research exploring teacher preparation for the digital age. *Cambridge Journal of Education*, *50*(1), 37–56. https://doi.org/10.1080/0305764X.2019.1625867.

Tondeur, J., Howard, S. K., & Yang, J. (2021). One-size does not fit all: Towards an adaptive model to develop preservice teachers' digital competencies. *Computers in Human Behavior*, *116*, 1–9. https://doi.org/10.1016/j.chb.2020.106659.

UNESCO. (2019). *International conference on Artificial Intelligence and education planning education in the AI Era: Lead the leap*. Beijing: Condpcept Note. Available at: https://en.unesco.org/sites/default/files/ai-conference-beijing-concept-note-en.pdf (Accessed April 2021).

Warschauer, M. (2003). *Technology and social inclusion: Rethinking the digital divide*. Cambridge, MA: MIT Press.

World Economic Forum. (2017). *Accelerating workforce reskilling for the fourth industrial revolution an agenda for leaders to shape the future of education, gender and work*. Available at: http://www3.weforum.org/docs/WEF_EGW_White_Paper_Reskilling.pdf (Accessed April 2021).

World Economic Forum. (2018). *The future jobs report 2018*. Available at: www.weforum.org/reports/the-future-of-jobs-report-2018 (Accessed April 2021).

World Economic Forum. (2020a). *The schools of the future. Defining new models of education for the fourth industrial revolution*. Available at: https://www3.weforum.org/docs/WEF_Schools_of_the_Future_Report_2019.pdf (Accessed March 2021).

World Economic Forum. (2020b). *The future jobs report*, *2020*. Available at: http://www3.weforum.org/docs/WEF_Future_of_Jobs_2020.pdf (Accessed April 2021).

Yeo, Y., & Lee, J.-D. (2020). Revitalizing the race *between* technology and education: Investigating the growth strategy for the knowledge-based economy based on a CGE analysis. *Technology in Society*, *62*(2020), 101295. ISSN 0160–791X. https://doi.org/10.1016/j.techsoc.2020.101295.

4 Chatbots in Education
A Systematic Review of the Science Literature

Antonio-José Moreno-Guerrero, José-Antonio Marín-Marín, Pablo Dúo-Terrón and Jesús López-Belmonte

CONTENTS

4.1 INTRODUCTION: BACKGROUND AND DRIVING FORCES

A chatbot is considered a computer programme that is able to hold a conversation with a human person, making use of technology based on Artificial Intelligence software (Touimi et al., 2020). This type of resource is integrated in various types of programmes that are commonly used on a daily basis. An example of this is Facebook or Telegram (Shumanov & Johnson, 2021). These platforms provide chatbots that try to answer various questions automatically, responding to the needs of the interlocutor (Rajaobelina & Ricard, 2021). A chatbot can be considered a powerful and useful tool, since it shares information and solves doubts (Abd-Alrazaq et al., 2021).

Among the characteristics that stand out in chatbots are agility of response and ability to learn (Tsai et al., 2021). That is, they are able to respond to interlocutors' queries quickly and concisely. Moreover, they are able to learn while interacting with people (Flanagan & Walker, 2020). They adapt their responses to the real needs of the interlocutors. This is because their foundations are based on Artificial Intelligence—specifically, machine learning, big data and natural language processing (Wang et al., 2020).

Among the actions that allow chatbots to be developed in education are the accessibility of website information, the simplification of administrative procedures, familiarizing members of the educational community with the educational culture itself, and the development of technical support and assistance for problem solving (Quiroga et al., 2020).

There are several companies involved in their development and manufacture, including Amazon, Apple and Microsoft. This indicates the great potential they can generate, given that they are backed by technological giants (Smutny & Schreiberova, 2020).

The use of chatbots is widespread in customer services, but in recent years it is generating interest in education (Yin et al., 2020). At the educational level it is relevant for parents, students and teachers. What is important in this type of resource is that it serves to maintain fluid and continuous communication with the interlocutors (Gros et al., 2020).

Unesco, OECD, Horizon and Forbes are examples of world organizations and reports that point to Artificial Intelligence as a priority to be incorporated into teaching practice (CODEIntef, 2019). These implementations can be useful to help teachers and students solve both educational questions and routine tasks (Chocarro et al., 2021). The use of AI offers new and prosperous opportunities given the increase in jobs that require STEM training (Rodríguez-García et al., 2020a).

For students, chatbots can be used to locate information and resolve doubts quickly, at any time of the day and automatically. They can also be used to book tutorials (Alm & Nkomo, 2020).

For teachers, chatbots can be used to monitor students' academic progress and development. They can also be used as a resource to support learning (Pereira et al., 2019).

For parents, chatbots can be used to maintain contact with schools, receiving advice or resolving doubts (Palasundram et al., 2019).

Regardless of the member of the educational community, chatbots should be presented as a simple and easy-to-use tool, given that there is a great diversity of digital competence among the people who can use them. Not all students, parents or teachers have similar levels of digital competence (Chen et al., 2020).

We highlight the article "LearningML: A Tool to Foster Computational Thinking Skills Through Practical Artificial Intelligence Projects" (Rodríguez-García et al., 2020a) with platforms that use AI and chatbot programming, from beginner levels in primary school, such as Machine Learning for Kids4 (ML4K) that can be exported to educational programming programs and initiation to Artificial Intelligence and creation of chatbots such as Scratch, Cospaces, LearningML.org, Code.org, MIT App Inventor or Python, highlighting that they are basic resources to get started in AI and Machine Learning, but insufficient to carry out complete projects.

Chatbots have a number of benefits for the educational environment, including (Abu, 2011; Pin-Chuan & Chang, 2020; Tamayo et al., 2020):

- They adapt to the needs of each student, adjusting to the pace of teaching and learning.
- They reinforce learning through repetition of information. People ask questions and the device responds appropriately to the question posed. This procedure can be repeated as many times as necessary.
- They promote spaced interval learning. That is, they remind learners of the lessons attended to so far so that they do not get forgotten. Repetition is key to memorizing certain content.

- They allow a student to be interviewed or questioned. This is developed if set up for this purpose.
- They can evaluate the actions or tasks elaborated by students.

There are a large number of resources that can be used to design chatbots in education, including CourseQ. This type of resource can establish a dialogue with students on issues such as class schedules or complementary activities. In addition, it becomes a support resource for teachers, as it allows them to answer students' questions and monitor their learning. All of this is automated (Lee et al., 2020).

It should be borne in mind that chatbots can be used for the most repetitive tasks in training processes, where they can answer students' theoretical questions. In this case, teachers, by not dedicating themselves to answering students' questions, have more time to develop more active and interactive dynamics in the training process (Gratzer & Goldbloom, 2020).

4.2 JUSTIFICATION AND APPLIED RESEARCH METHOD

Studies on Chatbots in the field of education date back to 2007. From that date to the present day, their production has been very irregular and scarce, with no more than 40 manuscripts on the subject. For this reason, and given the great relevance that Artificial Intelligence is acquiring in recent times (Pin-Chuan & Chang, 2020), a study of a bibliometric nature (Carmona-Serrano et al., 2021; Marín-Marín et al., 2021) is proposed, specifically a systematic literature review study (Rodríguez-García et al., 2020b).

This book chapter aims to determine the most relevant aspects of the scientific production on Chatbots in the educational field collected in the Web of Science (WoS) and Scopus databases. Both databases are considered highly relevant for research in the field of social sciences, as well as being the citation reference for Journal Citation Reports (JCR) and Scimago Journal & Country Rank (SJR).

The data collection has been developed by means of a search equation in the WoS and Scopus databases. The search was performed on the title of the manuscripts, as the term chatbots is considered a primary term. The search equation applied was as follows: "chatbot*". The asterisk was applied since the search was for both singular and plural manuscripts. In addition, the search was carried out in the main collection of Web of Science, specifically the following: SCI-EXPANDED, SSCI, A&HCI, CPCI-S, CPCI-SSH, BKCI-S, BKCI-SSH, ESCI, CCR-EXPANDED and IC, and in the main collection of Scopus.

For the proper development of this research, a PRISMA protocol has been applied. An initial search of 44 manuscripts in WoS and 136 manuscripts in Scopus was obtained, of which 22 were finally selected. Nine manuscripts that were not taken into account were eliminated because they belonged to the year 2021. They have been eliminated as this year has not yet been completed, to avoid bias in the research. For the selection of the remaining 22 documents, all types of manuscripts currently existing in WoS and Scopus have been considered. It should be noted that the final analysis of 22 manuscripts is due to the repetition of manuscripts in both databases.

The variables taken into account for the analysis of the manuscripts were the country, the date of publication, the main objectives of the research, the research methodology developed, the sample selected for the study, the variables applied in the research and the main findings.

4.3 RESULTS OF THE SYSTEMATIC LITERATURE REVIEW

The results achieved in the present study are presented to follow, after the analysis of the different variables. Table 4.1 shows all the articles reported from WoS and in Table 4.2 all the articles extracted from Scopus.

TABLE 4.1

Analysis of the Literature on Chatbots in Education in WoS.

Article 1	
Authors	Tamayo, P.A., Herrero, A., Martin, J., Navarro, C., Tranchez, J.M.
Year	2020
Country	Norway
Objectives	The work presents the reasons that motivated its adoption, the process of its development, differentiating two phases, its characteristics and functions, the assessment of its usefulness and the role of teachers in the implementation of this type of technological innovation.
Method	
Sample	
Variables	
Results	
Doi/URL	10.5944/openpraxis.12.1.1063
Article 2	
Authors	Lin, MPC Chang, D
Year	2020
Country	Taiwan
Objectives	In the present study, we developed a chatbot that helps teachers deliver written instructions.
Method	A preliminary analysis is made of the effect of a chatbot on the writing achievement of these writers. We also collect various testimonials from students about their chatbot experiences. Several important pedagogical and research implications have been addressed for chatbot-guided writing instructions and the use of learning technology.
Sample	
Variables	
Results	By working with the chatbot, the post-secondary writers developed a thesis statement for their argumentative essay outlines, and the chatbot helped the writers refine their peer-review comments.
Doi/URL	www.jstor.org/stable/26915408?seq=1

Article 3	
Authors	Salvat, B.G., Roig, A.E., Sanchez, M.P.
Year	2020
Country	Spain
Objectives	To know, through the co-design methodology, how mobile technologies can support migrant people.
Method	The data was obtained by means of a diagnosis through five analysis workshops and knowledge of the context, and five elaborations of scenarios carried out in Barcelona in the social entities that are part of the research team.
Sample	86 migrant persons participated (38 women and 48 men) from different countries, ages and educational levels, and who had been living in Barcelona for more than a year.
Variables	
Results	The results show a high use of smartphones and a generalized knowledge of the applications that are used mainly for communication. In addition, the experiences of migrants indicate that they can use existing tools to access information, but that they need additional support to take full advantage of the mobile tools available.
Doi/URL	https://idus.us.es/handle/11441/93682
Article 4	
Authors	Vasilateanu, A., Turcus, A.G.
	2019
Country	Spain
Objectives	Propose a system based on proactive chatbots for continuous and personalized mobile learning.
Method	
Sample	
Variables	
Results	• Learn about the PC's file system.
	• More information about the configuration of the PC.
	• More information about Windows applications.
	• Dive into your security system.
	• Get more information about networks and Internet access.
Doi/URL	https://library.iated.org/view/VASILATEANU2019CHA
Article 5	
Authors	Gutiérrez, E. Baldassarre, M Boticario, JG
Year	2019
Country	Spain
Objectives	Answer questions about the accessibility of chatbots, conversational agents or the topic of the virtual assistant.
Method	
Sample	
Variables	
Results	
Doi/URL	https://library.iated.org/view/GUTIERREZYRESTREPO2019ACC

(Continued)

TABLE 4.1 *(Continued)*

Article 6	
Authors	Samyn, K.
Year	2019
Country	Spain
Objectives	Develop a chatbot solution that uses a 3D chatbot to help with student training.
Method	
Sample	
Variables	
Results	
Doi/URL	https://library.iated.org/view/SAMYN20193DC
Article 7	
Authors	Saez-Fernandez, MD
	Escobar, P
	Marco-Such, M
	Candela, G
Year	2019
Country	Spain
Objectives	The application of chatbots in teaching would facilitate the work of teachers and provide immediate answers to frequently asked questions and virtual tutorials of students.
Method	Through the analysis of the tutorials, the different instances of conversational flows have been designed; the formative sentences have been extracted and classified according to the type of tutorial; finally, the different intentions have been established.
Sample	The chatbot prototype that is presented in this work has been formed through a set of tutorials provided by several professors from different subjects of the Computer Engineering and Multimedia Engineering fields.
Variables	
Results	Tutorials are a rich source of information for training an educational chatbot. The analysis of the text of the tutorials shows that in a high percentage the same questions are repeated by different students.
Doi/URL	10.21125/inted.2019.2175

TABLE 4.2
Analysis of the literature on chatbots in education in Scopus

Article 1	
Authors	Nawaz, N., Saldeen, M.A.
Year	2020
Country	Bahrain
Objectives	Identify the potential of applying Artificial Intelligence chatbot applications to library reference services.
Method	This document systematically reviews and explores AI chatbot embedded applications in academic library reference services.
Sample	

Variables	
Results	The researchers further explored that more flexibility and quality analysis can be received from a chatbot development framework.
Doi/URL	https://cutt.ly/bmeAxn3
Article 2	
Authors	Arias-Navarrete, A.S., Palacios-Pacheco, X.I., Villegas-Ch., W.
Year	2020
Country	Ecuador
Objectives	Implementing a learning environment based on student needs requires a great effort to encompass the large number of variables involved in education.
Method	Identification of variables. Analysis of data. Recommending system. Integration of LMS and Artificial Intelligence.
Sample	
Variables	
Results	
Doi/URL	www.proquest.com/docview/2452331734
Article 3	
Authors	Stathakarou, N., Nifakos, S., Karlgren, K., Konstantinidis, S.T., Bamidis, P.D., Pattichis, C.S., Davoody, N.
Year	2020
Country	Cyprus
Objectives	The objective of this course is to provide students with tools and methods to analyze and model the needs and requirements of patients, health professionals and care providers, as well as to evaluate health systems in different contexts.
Method	Thematic content analysis on the results of the students' tasks.
Sample	
Variables	
Results	
Doi/URL	https://ebooks.iospress.nl/publication/54631
Article 4	
Authors	Al-Ghadhban, D., Al-Twairesh, N.
Year	2020
Country	Saudi Arabia
Objectives	In this study, we have developed "Nabiha", a chatbot that can support conversation with information technology (IT) students at King Saud University using the Saudi Arabic dialect.
Method	
Sample	
Variables	
Results	
Doi/URL	https://thesai.org/Publications/ViewPaper?Volume=11&Issue=3&Code=IJACSA&SerialNo=57

(Continued)

TABLE 4.2 *(Continued)*

Article 5

Authors	Mellado-Silva, R., Faúndez-Ugalde, A., Blanco-Lobos, M.
Year	2020
Country	Chile
Objectives	This article will show the experience resulting from the use of a chatbot to support learning in accounting students for the teaching of tax regulations related to the Chilean tax system.
Method	Two types of tools are compared, on the one hand, a free-talking chatbot using natural language processing versus a rule-based chatbot driven by a decision tree.
Sample	The experimentation process was carried out with 50 higher education students, divided into an experimental group and a control group, in two different courses.
Variables	
Results	The results obtained demonstrated in both cases a greater effectiveness of the use of the chatbot in learning the tax matter, both in the free conversation chatbot where the experimental group obtained an improvement of 15.7% compared to the control group that obtained an improvement of the 1.05%, as in the chatbot that applied decision tree where the experimental group obtained an improvement of 32% versus the control group with 5.2%.
Doi/URL	https://astesj.com/v05/i06/p52/

Article 6

Authors	Wu, E.H.-K., Lin, C.-H., Ou, Y.-Y., Liu, C.-Z., Wang, W.-K., Chao, C.-Y.
Year	2020
Country	Taiwan
Objectives	Build a hybrid model chatbot learning assistant and explore its potential to improve the user experience in E-Learning and reduce users' sense of isolation and detachment while increasing their learning motivation.
Method	A chatbot is compared to a teacher advisory service on the E-Learning platform on which our chatbot is based to see how well our chatbot performs.
Sample	
Variables	
Results	The results of the evaluation of the experiment and the questionnaire show that chatbots could be useful for learning and could potentially reduce the feelings of isolation and detachment of E-Learning users.
Doi/URL	https://ieeexplore.ieee.org/document/9069183?denied=

Article 7

Authors	Mendez, S.L., Johanson, K., Conley, V.M., Gosha, K., Mack, N., Haynes, C., Gerhardt, R.
Year	2020
Country	USA
Objectives	The purpose of this paper is to explore the effectiveness of simulated interactive virtual conversations (chatbots) for mentoring underrepresented minority engineering PhD students who are considering pursuing a career in teaching or in industry.
Method	The effectiveness of the chatbot is examined through a phenomenological design.
Sample	Focus groups are used with underrepresented minority doctoral engineering students.

Variables	The four-stage process of phenomenological data analysis was followed: epoché, horizontalization, imaginative variation, and synthesis.
Results	While underrepresented minority PhD engineering students have extensive unmet tutoring needs and are generally satisfied with the user interface and reliability of chatbots, their ability to use them is mixed due to lack of customization in this type of complementary mentoring relationship.
Doi/URL	www.informingscience.org/Publications/4579

Article 8

Authors	Palasundram, K., Sharef, N.M., Nasharuddin, N.A., Kasmiran, K.A., Azman, A.
Year	2019
Country	Malaysia
Objectives	
Method	
Sample	
Variables	
Results	The generated response is compared with the gold response. The BLEU score is between 0 and 1, where 0 means no match at all and 1 means perfect match and anything in between means there are some overlaps between the two texts. For questions that have a one-to-many mapping, one of the answers is accepted as the correct answer. Models with the highest BLEU score are considered the best.
Doi/URL	https://online-journals.org/index.php/i-jet/article/view/12187

Article 9

Authors	Priadko, A.O., Osadcha, K.P., Kruhlyk, V.S., Rakovych, V.A.
Year	2019
Country	Ukraine
Objectives	Describe the process of developing a chatbot to provide students with timetable information using the Telegram mobile messenger.
Method	
Sample	
Variables	
Results	
Doi/URL	https://cutt.ly/umeAc5g

Article 10

Authors	Choi, S., Seo, W., Choi, S., Jang, E.-K.
Year	2018
Country	South Korea
Objectives	This study presents data and a framework for developing a chatbot for security education.
Method	
Sample	
Variables	
Results	
Doi/URL	www.icicelb.org/ellb/contents/2018/11/elb-09-11-06.pdf

(Continued)

TABLE 4.2 *(Continued)*

Article 11

Authors	Cliffe Schreuders, Z., Shaw, T., Muireadhaigh, A.M., Staniforth, P.
Year	2018
Country	United Kingdom
Objectives	Create a new approach to cyber-security. Create reusable random challenges and scenarios that could be used for learning and assessment purposes.
Method	A design science research approach was applied, by designing a solution, followed by implementation and evaluation.
Sample	
Variables	
Results	The results were encouraging, finding the approach convenient, attractive, fun and interactive; while significantly reducing manual dialing workload for staff.
Doi/URL	www.usenix.org/system/files/conference/ase18/ase18-paper_schreuders.pdf

Article 12

Authors	Carayannopoulos, S.
Year	2018
Country	Canada
Objectives	The purpose of this document is to examine how chatbots can be used to address two key problems faced by students in the first year: the feeling of being disconnected from the instructor and information overload.
Method	A tool was designed, implemented and tested against research insights.
Sample	
Variables	
Results	The document reveals the results of the application of this tool in a large freshman class and proposes improvements for future interactions.
Doi/URL	www.emerald.com/insight/content/doi/10.1108/IJILT-10-2017-0097/full/html

Article 13

Authors	Griol, D., Callejas, Z.
Year	2013
Country	Spain
Objectives	In this article we present an architecture that facilitates the construction of interactive pedagogical chatbots that can interact with students in natural language.
Method	
Sample	
Variables	
Results	The study showed a high degree of satisfaction with the appearance and interface of the system, and the results were very positive regarding its pedagogic potential.
Doi/URL	https://journals.sagepub.com/doi/full/10.5772/55791

Article 14

Authors	Crown, S., Fuentes, A., Jones, R., Nambiar, R., Crown, D.
Year	2010

Country	USA
Objectives	The purpose of this interactive online learning environment is to encourage students to think reflectively about the fundamentals of the course.
Method	
Sample	
Variables	
Results	
Doi/URL	https://cutt.ly/bmeAbIp
Article 15	
Authors	Kerly, A., Hall, P., Bull, S.
Year	2007
Country	United Kingdom
Objectives	This document analyzes the development and capabilities of conversational agents (or chatbots) and intelligent tutoring systems, in particular open learning modeling.
Method	We describe a Wizard of Oz experiment to investigate the feasibility of using a chatbot to support negotiation and conclude that a merger of the two fields can lead to the development of negotiation techniques for chatbots and the improvement of the open learning model.
Sample	
Variables	
Results	
Doi/URL	www.sciencedirect.com/science/article/abs/pii/S0950705106001912

4.4 DISCUSSION AND CONCLUSIONS

After analyzing the documents reported from both databases, it is revealed that the literature on chatbots in the field of education is presented as a field of study yet to be discovered in its entirety. The scarcity of articles based on Artificial Intelligence and the creation of chatbots in education allows us to reflect on its scope in the classroom as a methodology, in addition to ethical issues.

Designing chatbots and the different communication methods used improves the educational community's vision for their employment (Chocarro et al., 2021) from teaching and/or learning, although there should be a greater number of articles based on experiences. The construction of a chatbot as a methodological resource in education allows students to learn material thoroughly and better understand the world around them.

Rodríguez-García et al. (2020a) affirm that introducing practical projects based on AI is the way to educate citizens and develop critical thinking and awaken interest among young people in the field. However, the small number of successful experiences published do not contribute to teachers and students becoming creators and daily users of chatbots through Artificial Intelligence and the development of machine learning, as CODEIntef points out in 2019.

The number of publications in article format on the state of affairs is small. Only seven manuscripts have been reported in the WoS database, while 15 articles on the subject have been extracted from the Scopus database.

Regarding the years of publication, most have been published in the last three years, although articles have been reported since 2007. This indicates that chatbots in the educational field are not a phenomenon that has just appeared but that for several years the state of the issue has been investigated. However, the concept is currently being projected in the impact literature.

All this is mainly due to the inclusion of technology in the field of education (López-Belmonte et al., 2020). For this reason, chatbots intend to be integrated into the educational field to favor training and tutoring practices. This will facilitate the improvement of teaching and learning processes, to achieve the adaptation of the training actions to a society marked by the incidence of technology in each and every one of the daily actions (Soler-Costa et al., 2021).

The prospective of this work is focused on publicizing articles about chatbot in education. In this way, readers and researchers interested in the subject will have a source where they can go to expand their knowledge. On the other hand, this chapter reflects the lack of studies on the construct in question. Therefore, we encourage the scientific community to investigate and focus its efforts on providing new findings and evidence of this technology steadily making its way into education.

The limitations of this work have been several. Many of the documents reported from the aforementioned databases were not open access. Also, some of the variables analyzed have not been collected in the reported manuscripts. As a future line of study, it is intended to carry out an educational experience through chatbots for the tutoring of students in various educational stages.

REFERENCES

Abd-Alrazaq, A. A., Alajlani, M., Ali, N., Denecke, K., Bewick, B. M., & Househ, M. (2021). Perceptions and opinions of patients about mental health chatbots: Scoping review. *Journal of Medical Internet Research*, *23*(1), 1–16. doi: 10.2196/17828.

Abu, B. (2011). A chatbot as a natural web interface to Arabic web QA. *International Journal of Emerging Technologies in Learning*, *6*(1), 37–43. doi: 10.3991/ijet.v6i1.1502.

Alm, A., & Nkomo, L. M. (2020). Chatbot experiences of informal language learners: A sentiment analysis. *International Journal of Computer-Assisted Language Learning and Teaching*, *10*(4), 51–65. doi: 10.4018/IJCALLT.2020100104.

Carmona-Serrano, N., Moreno-Guerrero, A.-J., Marín-Marín, J.-A., & López-Belmonte, J. (2021). Evolution of the autism literature and the influence of parents: A scientific mapping in web of science. *Brain Sciences*, *11*(1), 1–16. doi: 10.3390/brainsci11010074.

Chen, H. L., Widarso, G. V., & Sutrisno, H. (2020). A ChatBot for learning Chinese: Learning achievement and technology acceptance. *Journal of Educational Computing Research*, *58*(6), 1161–1189. doi: 10.1177/0735633120929622.

Chocarro, R., Cortiñas, M., & Matás, G. M. (2021). Teachers' attitudes towards chatbots in education: A technology acceptance model approach considering the effect of social language, bot proactiveness, and users' characteristics. *Educational Studies*, 1–19. doi: 10.1080/03055698.2020.1850426.

Flanagan, F., & Walker, M. (2020). How can unions use Artificial Intelligence to build power? The use of AI chatbots for labour organising in the US and Australia. *New Technology Work and Employment*, 1–16. doi: 10.1111/ntwe.12178.

Gratzer, D., & Goldbloom, D. (2020). Therapy and E-therapy-preparing future psychiatrists in the era of apps and chatbots. *Academic Psychiatry*, *44*(2), 231–234. doi: 10.1007/s40596-019-01170-3.

Gros, B., Escofet, A., & Paya, M. (2020). Co-design of a chatbot to facilitate administrative procedures for migrants. *Pixel Bit. Revista de Medios y Educación*, *57*, 91–106.

Lee, J. H., Yang, H., Shin, D., & Kim, H. (2020). Chatbots. *Elt Journal*, *74*(3), 338–344.

López-Belmonte, J., Marín-Marín, J. A., Soler-Costa, R., & Moreno-Guerrero, A. J. (2020). Arduino advances in web of science. A scientific mapping of literary production. *IEEE Access*, *8*, 128674–128682. doi: 10.1109/ACCESS.2020.3008572.

Marín-Marín, J. A., Moreno-Guerrero, A. J., Dúo-Terrón, P., & López-Belmonte, J. (2021). STEAM in education: A bibliometric analysis of performance and co-words in web of science. *International Journal of STEM Education*, *8*, 1–21. doi: 10.1186/s40594-021-00296-x.

Ministerio de Educación y Formación Profesional. (2019). Inteligencia artificial en el aula con Scratch 3.0. *CODEIntef*. http://code.intef.es/inteligencia-artificial-en-el-aula-con-scratch-3–0/.

Palasundram, K., Sharef, N. M., Nasharuddin, N. A., Kasmiran, K. A., & Azman, A. (2019). Sequence to sequence model performance for education chatbot. *International Journal of Emerging Technologies in Learning*, *14*(24), 56–68. doi: 10.3991/ijet.v14i24.12187.

Pereira, J., Fernández-Raga, M., Osuna-Acedo, S., Roura-Redondo, M., Almazan-López, O., & Buldon-Olalla, A. (2019). Promoting learners' voice productions using chatbots as a tool for improving the learning process in a MOOC. *Technology Knowledge and Learning*, *24*(4), 545–565. doi: 10.1007/s10758-019-09414-9.

Pin-Chuan, M., & Chang, D. (2020). Enhancing post-secondary writers' writing skills with a chatbot: A mixed-method classroom study. *Educational Technology & Society*, *23*(1), 78–92.

Quiroga, J., Daradoumis, T., & Marqués, J. M. (2020). Rediscovering the use of chatbots in education: A systematic literature review. *Computer Applications in Engineering Education*, *28*(6), 1549–1565. doi: 10.1002/cae.22326.

Rajaobelina, L., & Ricard, L. (2021). Classifying potential users of live chat services and chatbots. *Journal of Financial Services Marketing*, 1–11. doi: 10.1057/s41264-021-00086-0.

Rodríguez-García, A.-M., Moreno-Guerrero, A.-J., & López Belmonte, J. (2020a). Nomophobia: An individual's growing fear of being without a smartphone—A systematic literature review. *International Journal of Environmental Research and Public Health*, *17*(2), 1–19. doi: 10.3390/ijerph17020580.

Rodríguez-García, J. D., Moreno-León, J., Román-González, M., & Robles, G. (2020b). LearningML: A tool to foster computational thinking skills through practical Artificial Intelligence projects. RED. *Revista Educación a Distancia*, *20*(63). http://dx.doi.org/10.6018/red.410121

Shumanov, M., & Johnson, L. (2021). Making conversations with chatbots more personalized. *Computer in Human Behavior*, *117*, 1–14. doi: 10.1016/j.chb.2020.106627.

Smutny, P., & Schreiberova, P. (2020). Chatbots for learning: A review of educational chatbots for the Facebook Messenger. *Computer & Education*, *151*, 1–11. doi: 10.1016/j.compedu.2020.103862.

Soler-Costa, R., Moreno-Guerrero, A.-J., López-Belmonte, J., & Marín-Marín, J.-A. (2021). Co-word analysis and academic performance of the term TPACK in web of science. *Sustainability*, *13*(3), 1–20. doi: 10.3390/su13031481.

Tamayo, P. A., Herrero, A., Martín, J., Navarro, C., & Tranchez, J. M. (2020). Design of a chatbot as a distance-learning assistant. *Open Praxis*, *12*(1), 145–153. doi: 10.5944/openpraxis.12.1.1063.

Touimi, Y. B., Hadioui, A., El Faddouli, N., & Bennani, S. (2020). Intelligent chatbot-LDA recommender system. *International Journal of Emerging Technologies in Learning*, *15*(20), 4–20. doi: 10.3991/ijet.v15i20.15657.

Tsai, M. H., Chan, H. Y., Chan, Y. L., Shen, H. K., Lin, P. Y., & Hsu, C. W. (2021). A chatbot system to support mine safety procedures during natural disasters. *Sustainability*, *13*(2), 1–13. doi: 10.3390/su13020654.

Wang, Y., Zhang, N., & Zhao, X. (2020). Understanding the determinants in the different government AI adoption stages: Evidence of local government chatbots in China. *Social Science Computer Review*, 1–13. doi: 10.1177/0894439320980132.

Yin, J. Q., Goh, T. T., Yang, B., & Xiaobin, Y. (2020). Conversation technology with micro-learning: The impact of chatbot-based learning on students' learning motivation and performance. *Journal of Educational Computing Research*, *59*(1), 154–177. doi: 10.1177/0735633120952067.

5 Applications of Artificial Intelligence in Learning Assessment

Trishul Kulkarni, Bhagwan Toksha and Prashant Gupta

CONTENTS

5.1 INTRODUCTION

The design of any educational system starts with the development of learning objectives, instructional content, assessment methods, pedagogical procedures and associated strategies. The most important part is the assessment and evaluation system to determine whether learning objectives are achieved or not by learners. These learning goals can be classified into three learning domains: cognitive, affective and psychomotor. The cognitive domain entails the know-how and the improvement of intellectual skills; the affective domain of learning is concerned with how emotions, feelings, and attitude affect students' learning; and the psychomotor domain is concerned with bodily movement, coordination and use of motor-ability skills. These three domains appear to be independent of each other. However, with a viewpoint of a holistic approach, all are interconnected and they overlap with each other as shown in Figure 5.1. The cognitive assessment is useful to evaluate the ability of the student in the comprehension of knowledge and intellectual skills; affective assessment is required to evaluate the emotional response of the student while accepting the knowledge and skills; whereas the psychomotor assessment is essential to evaluate how well a student is implementing the knowledge and skills with respect to the motor ability.

The traditional learning assessment is mainly focused on the assessment of cognitive skills. The cognitive competencies receive more emphasis as automation is easy

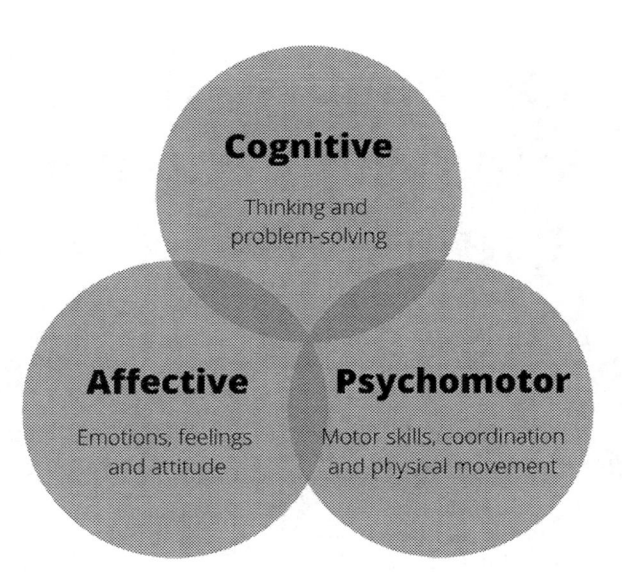

FIGURE 5.1 Three Major Domains of Learning as Cognitive, Affective and Psychomotor, with Their Features.

to do for the cognitive assessment using ICT tools. On the other hand, it is difficult to automate the assessment of affective and psychomotor competencies with traditional ICT tools. However, this automation can be used in a limited sense for the assessment of lower-level cognitive skills.

Progress in the field of ICT has made it possible to learn anywhere and anytime for everyone. Artificial Intelligence (AI) has added a new dimension to these ICT developments. Moreover, in the recent times we are witnessing advancements in various AI techniques such as Machine Learning (ML), Deep Learning (DL), Image Processing (IP), Natural Language Processing (NLP) and Speech Recognition. These techniques have potential to change various aspects of the educational system. AI could take automation in the educational fields to a next level, and could change the way we find and use educational information. Also, it can bring changes in the role of a teacher, the way we assess the students and provide feedback (Braiki et al., 2020; Gardner et al., 2021; Zhai et al., 2021).

AI techniques take inspiration from the human brain functions in a way, as ML and DL could resemble thinking and analytical skills, with IP resembling the eye and brain coordination and NLP performing the task involved in the reasoning and understanding of natural language. With the help of various machine learning algorithms, AI can identify the patterns in a dataset and performing predictive tasks without any need to explicitly programme a computer for that specific task. There are broadly two types of machine learning algorithms—supervised and unsupervised algorithms. Supervised machine learning algorithms make the use of a training dataset which contains both input and output data points for a specific system. These algorithms learn the pattern associated with a dataset in a similar way a human brain will work to learn a system. Based on this learning, these algorithms can predict the output for unknown data points. On the other hand, unsupervised machine learning algorithms identify the hidden patterns in a non-labeled dataset so that it can be analyzed without any need of supervision (Jain et al., 2000). Deep learning is a subset of machine learning techniques inspired by the neural structure of the human brain. IP is a technique of AI which deals with data extraction and classification performed on images or computer vision. NLP is an important sub-domain of AI which makes computers capable of interpreting the human language. All these AI techniques can be applied for the assessment of students' progress and performance in all three learning domains. In other words, computer systems can acquire the capacity to judge students' performance in the same manner as the functioning of a human brain.

There is growing interest in using AI for formative assessment (Gardner et al., 2021; Vittorini et al., 2020). Formative assessment is carried out during the sessions rather than at the end for providing feedback to the student. It is more of 'assessment for learning' rather than 'assessment of learning'. Formative assessment can be effective only if it is capable of providing immediate feedback to students about their learning. Manual scoring of assessment items is a time consuming and skill-oriented task. It would be unrealistic to expect the immediate feedback with the manual assessment methods. AI-based assessment should not be thought of only as automation of assessment but as a system to achieve an immediate intelligent feedback mechanism. This mechanism will be of great assistance for the learner to regulate their own learning progress (Cicchinelli et al., 2018).

The prediction of students' performance at the early stage of an academic session is one of the most promising applications of AI in the educational system. This type of prediction can identify students who are under-performing and are at the risk of being dropped out. Academic monitoring systems can execute the corrective actions to improve the performance of such students and reduce the dropout rate (Arora et al., 2014; Manhães et al., 2014). Educational institutions generate large amounts of educational data regarding students' progress. This data typically includes evaluations based on formative and summative assessment. Educational data mining (EDM) techniques allow academicians to build an academic monitoring system to predict the student performance based on similar data available with institutions. Various EDM techniques like Naive Bayes, instance-based learning classifier, Support Vector Machine, Decision Tree and Neural Network are employed to predict the student progress (Costa et al., 2017).

The present chapter is an overview and critique about the advancements of AI as a part of assessment systems in three domains of learning. The challenges faced in the conventional assessment, evaluations and the opportunities provided by the AI enabled assessment tools are discussed. Academicians need to understand the role of AI in an assessment mechanism for its effective inclusion. This chapter will help academicians to identify and implement the most appropriate AI tools in the respective context. The chapter is not intended to explore the technical details of AI—rather the aim is to make academicians familiar with the various possibilities by which AI can be incorporated in the assessment and evaluation part of education.

5.2 AI FOR ASSESSMENT IN COGNITIVE DOMAIN

The complete cognitive domain can be represented by various learning processes categorized at six levels as prescribed by Bloom's taxonomy (Bloom, 1956). These categories can be considered as difficulty levels to be mastered earlier before moving to the subsequent level. The processes related to recall or retrieval of specific information or facts are at the base (remembering). The learning processes in the second category are related to comprehension and understanding of concepts (understanding). The third category comprises of application of previously acquired facts or information in completely new situations (applying). The learning processes in the fourth category are related to the comparing, contrasting and drawing of relationships among different facts or concepts (analyzing). The learning processes in the fifth category are related to making judgments as a part decision-making process (evaluating) while the last category is related to creation of new meaning or structure (creating) (Anderson et al., 2001).

The learning assessment of students in the cognitive domain is mainly concerned with assessment of students' abilities in the comprehension of knowledge and various intellectual skills which are well received by the educational system in the form of Bloom's taxonomy as shown in Figure 5.2. There are established systems to validate the students' learning in the lower order thinking skills. However, the assessment and evaluation of these skills is still performed manually, which is time and resource consuming. Moreover, the motive of assessment gets defeated as it lacks the ability to provide immediate feedback to the students (Epstein et al., 2002).

Auto grading technology is widely employed in the various online educational environments such as Learning Management Systems (LMS) and Massive Open Online Courses (MOOCs). These grading techniques are mostly limited to multiple choice types of objective questions. It is quite easy to apply computational grading to typical multiple-choice types of questions as there are certain numbers of correct responses to each question. Automating the grading of open-ended questions like essays is also not a new idea. Page Ellis B. first proposed the idea of grading of essays

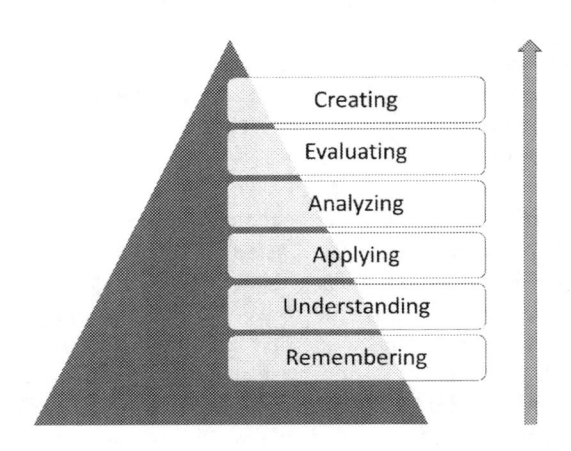

FIGURE 5.2 Bloom's Taxonomy for the Cognitive Domain.

by computational facilities in 1966 (Page, 1966). However, it is difficult to adopt this technique for open-ended or subjective questions, which demand a constructive response from students. This creates a requirement of training computers for the understanding of natural language (Leacock & Chodorow, 2003). In order to speed up the assessment of the subjective response, a convenient solution of 'calibrated peer review' is often employed. In this method, students review their peers' work anonymously and with a predefined rubric. Even though this system helps to reduce the burden of instructor, it hardly automates the assessment process (Robinson, 2001). The AI techniques such as ML and NLP are employed to automate the grading of subjective responses which involves a constructed response from the student. The basic idea is to use the supervised ML model which is trained by a known dataset to predict the outcome of an unseen test dataset. The 'training data' consist of sample assignments which are graded by experts as an input to the ML model. The ML programme would process that data and extract the features and patterns out of it to predict the assignment grades for ungraded assignments. This predictive capability of ML depends on several factors, such as

- How well the problem is formulated.
- Availability of training data set, generally bigger the data set better is the accuracy.
- Use of specific ML algorithms.
- How well the features have been identified.

The ML problem can either be formulated as, (Geigle et al., 2016)

1) Classification problem in which assignments are classified into number of groups, where each group represents a grade.
2) Ranking problem in which ungraded assignments are ranked based on quality without assigning any specific grades to it.

Several auto grading applications for the constructed response have been reported in the literature. Mitchell et al. developed software 'AutoMark', which is capable of comparing the students' responses for specific questions against standard mark scheme templates (Mitchell et al., 2002). C-rater is an auto grading programme which analyzes student response to an open-ended question with the help of NLP techniques and accordingly full, partial or no credit (marks) is assigned to a student response (Leacock & Chodorow, 2003). Haudek et al. investigated the use of IBM's SPSS text analysis to categorize the students' responses into lexical categories and thereby predicting the score (Haudek et al., 2012). Dzikovska et al. developed an application to categorize the students' responses in five different categories by making use of textual entailment and NLP (Dzikovska et al., 2013).

Most of the time, lower cognitive level assessment items are used in written or oral examinations targeting the recall and understanding levels. It is essential to include assessment items related to higher cognitive levels which help students to enhance higher order thinking skills (Vogler, 2005). The application of AI can help an educationalist in this regard. Yahya et al. reported an interesting application of machine learning to classify the test questions according to cognitive levels of Bloom's taxonomy. Such applications can help the LMS to identify and confirm the cognitive level of assessment items thereby enforcing the use of higher levels of cognitive assessment (Yahya et al., 2013).

In the current scenario, various MOOC platforms are using 'calibrated peer review' (CPR) activities as a part of course assessment. CPR techniques have been derived from peer reviewing of scientific literature in the scientific community. The student gets an opportunity to review work performed by peers by making use of a structured rubric prepared by the instructors, and benefits by analyzing the quality of work produced by peers. This kind of reviewing process involves the critical thinking skills and higher cognitive levels (Robinson, 2001). The conventional peer review processes as part of online educational platforms use computational power merely to store and retrieve the review data. Ramachandran et al. proposed an automatic peer review assessment system. This system can provide immediate feedback to the reviewer about his or her submission. The system generated feedback allocating the score for each review submitted based on the content, relevance and coverage using NLP and ML techniques (Ramachandran et al., 2017).

ACORN (Assessing Contextual Reasoning about Natural Selection) is an assessment tool designed to assess students' understanding about the phenomenon of natural selection in evolutionary biology. ACORN is used as a diagnostic test to expose and address students' misconceptions related to the subject by examining their understanding about the normative scientific idea and non-normative naïve ideas about evolutionary changes. In this diagnostic test, students are asked to write short answers to the questions addressing specific reasoning context. Multiple ACORNS items (i.e., questions) are formulated to expose the students reasoning about evolution and natural selection. However, the major limitation for using such an assessment tool is that it involves manual scoring of student responses which makes it difficult to get immediate feedback. The most important quality of an assessment system to be used for diagnostic or formative assessment is the capacity to provide instant feedback, and a corrective action can be taken based on that feedback. In order to mitigate this challenge, a ML-based assessment tool EvoGrader was developed for use in an introductory undergraduate biology class. This online tool is available in public domain; *www.evograder.org*. EvoGrader is designed to address 86 different open-ended questions (i.e., ACORNS items). The functioning of EvoGrader is divided in three different steps. The first step involves the students' writing answers for ACORNS assessment items. The instructor then uploads a. csv file containing responses to these questions in the form of short answers. In the second step, the system automatically grades and analyzes the students' response to the ACORNS items. The third and final step involves the grade report generation which also comprises how well the key concepts are grasped by each student on an individual basis. The system employs a supervised learning model for grading student responses based on key concepts, naïve idea and holistic reasoning model scores. The training and testing dataset for the supervised machine learning was made available in the form of pre-scored written explanations to each ACORNS item by the domain experts. The machine learning model classifies student responses based on presence or absence of key concepts in the answer text.

EvoGrader works with the help of two modules referred to as components, as shown in Figure 5.3, including a scoring module which is at front-end and a onetime training module at the back-end. The scoring module is an online component of the system supported by cloud services and its function is to receive the responses to the ACORNS items in. csv format, interact with the back-end module for predicting grades by classifying the given responses, and generate the interpretation and visualization of results to be presented to the end user (student). The training module at the back-end is an offline component processed at once. It is responsible for building a scoring model with the help of a training dataset for each concept, which can be used multiple times by the front-end module of the EvoGrader. The back-end module achieves the prediction of grades by a series of three functions. Firstly, it extracts the features from the training dataset by using NLP tools like "bag of words". The second stage is to construct a prediction model for which it makes use of Sequential Minimal Optimization (SMO) algorithm for training the Support Vector Machines. These vectors, which are also termed as "feature vectors", are nothing but the responses to ACORNS items which are converted into binary data depending upon the presence or absence of normative scientific idea and non-normative naïve ideas about the evolutionary changes in the written explanations. In order to train the scoring model at the back-end, the system uses more than ten thousand responses collected from around three thousand participants. These responses are validated and scored by the experts. The grades predicted by the EvoGrader are closely comparable with the grades awarded by human graders. Thus, the system can be used as an assessment tool with near perfection (Moharreri et al., 2014). Such instances are encouraging to design assessment tools with the help of AI. However, an exhaustive survey aimed at assessment and evaluation of higher order thinking in higher education could not

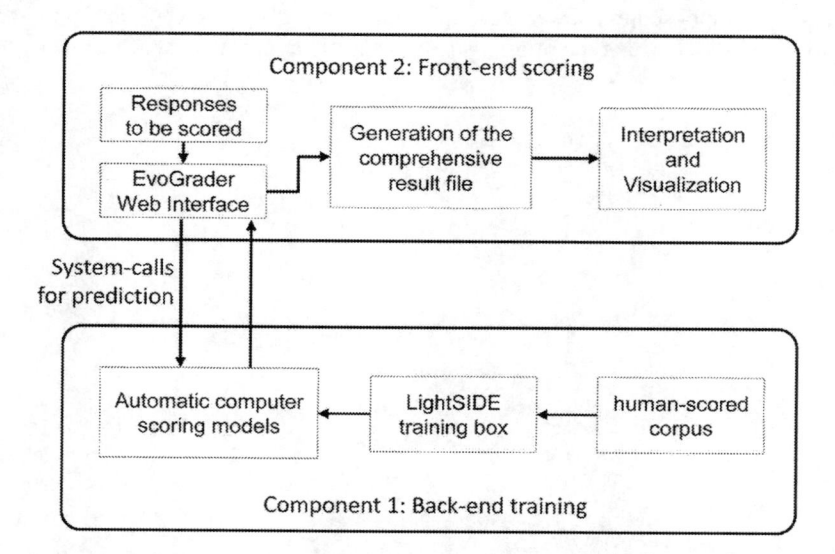

FIGURE 5.3 Architecture of Evograder with Front-End and Back-End Components (Moharreri et al., 2014).

locate any of such instances. This remains as a future scope for educationalists and technologists to collaborate and fetch better results in this direction.

5.3   AI FOR ASSESSMENT IN AFFECTIVE DOMAIN

The affective domain is one of the important domains out of the three Blooms taxonomy learning domains. It is concerned with how emotions, feelings and attitude affect students' learning. The students' development in the affective domain can be represented by various stages as shown in Figure 5.4. The initial stage is being attentive which indicates the willingness to receive any form of learning (receiving). It can be as simple as listening to instructor attentively. The second stage is about the engagement and active participation of learner in receiving the knowledge and skills (responding). The next stage is 'valuing' which refers to the ability to judge the worthiness of something and offer commitment towards it. The fourth stage is about applying these values to her or his own life earned at the earlier stage (organizing). The final stage of affective domain is about internalizing and acting consistently with the value system adopted at earlier stages (characterizing) (Kraftwohl et al., 1967).

In a typical classroom setup, assessment and evaluation of these affective stages is often neglected. The instructor uses various techniques like asking questions and creating the opportunities for discussion to keep students attentive, but most of the time there is a lack of formal structure. These activities are mostly performed to keep the decorum of the classroom environment (Holt & Hannon, 2006). When it comes to distant learning methods, it becomes more difficult to keep track of students' attentiveness and responsiveness. E-learning platforms have devised various techniques like reflection quizzes at regular intervals throughout sessions in which certain questions, mostly of multiple-choice nature, are displayed on the screen. The students are supposed to answer these questions correctly before moving to the next part of the session. This type of assessment is mostly a formative type of assessment

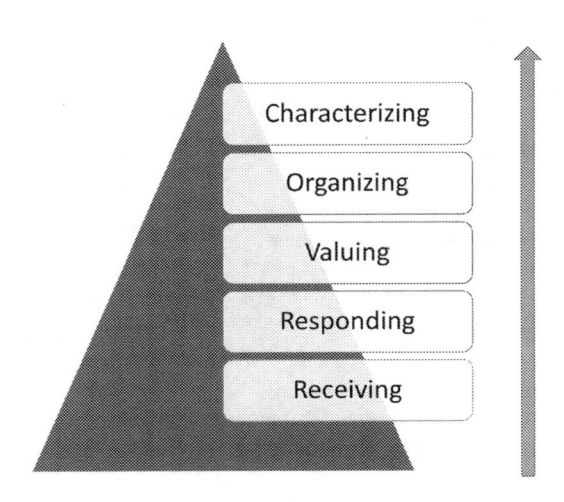

FIGURE 5.4   Hierarchy of the Affective Domain Related to Learning.

and is not counted towards final grades. The whole purpose of this type of assessment is to keep the students attentive and provide them an opportunity to respond. Another important tool used in online platforms is discussion boards or forums. Students are encouraged to participate in discussion forums and in some cases fractions of grades are awarded for meaningful participation in the discussion forum (Pena-Shaff & Nicholls, 2004). But it is very difficult for instructors to keep track of discussion boards or forums and evaluate the attentiveness of students. Also, these kinds of activities may not be suitable for all types of courses.

The assessment of attending and responding stages of affective domain can be achieved by monitoring student activities in synchronous mode with the help of IP and face recognition techniques. These systems track the facial movements of students through web cam and assign scores based on the status of student activities throughout the online session. The score which is a measure of students' attentiveness and responsiveness is computed by the application of fuzzy logic and a fuzzy integral. The attentiveness is measured through facial recognition and input device activities—mouse movement performed by the students during the online session—whereas responsiveness of students is computed by the level of participation in discussion forum (Yang, 2012). A small change in the facial expression of a learner (facial micro-expression state) while learning a difficult concept can be captured and used to evaluate whether he or she is able to understand the concept or not (Leonard Liaw et al., 2014). Though these systems capture the attentiveness of the learner by a measure of facial movements, it is a physical measurement, and the emotional state of learner is not addressed by such a system.

The field of affective computing has received a lot of interest in recent years, and it deals with the evaluation of human feelings and emotions captured with the aid of bodily sensors and affective algorithms (Zhai et al., 2021). It aims at enhancing the interaction among the human user and the machine by detecting the user emotions and feelings so that the machine or system can provide the appropriate response. It is possible to detect and classify even one of the primitive emotion like fear by using various physiological signals like electroencephalograph (EEG), different ML models and DL techniques (Bălan et al., 2019). Affective computing finds very interesting applications in the field of education and assessment. Researchers have developed an intelligent tutoring system (ITS) which can detect the user's emotional state by recognition of facial expressions and text semantics. When learners interact with ITS, the emotional and mental state of a learner can be accessed by measuring the brainwaves which are captured by wireless EEG devices. These devices can be used as necklaces or pendants by the students. This data can be used to adapt the different teaching strategies as per the learners' affective status and to provide appropriate feedback to the learner (Lin et al., 2012). Zhai et al. demonstrated that learners' reading comprehension was significantly improved by using a bio-feedback technique which utilizes physiological signals related to gazing, speech and brain activities captured through eye tracking and EEG devices (Zhai et al., 2018).

It is a herculean task for a teacher to assess the students' engagement in a typical classroom. However, quantitative measurement of students' behaviour in classrooms is possible with the help of students' engagement recognition system reported by Soloviev V. It is now possible to get the quantitative answers to questions such as how well students are engaged in learning. Are students enjoying the learning? Are students participating in activities? The engagement recognition system was deployed in Financial University under the Government of the Russian Federation as a pilot study. It makes use of IP and ML to assess students' engagement in the classroom. The facial images of students were captured from a continuous video stream with the help of cameras placed in the classroom. The external experts were employed to recognize the emotional state of a student to determine the level of engagement. These facial images were categorized in to two groups based on the level of engagement: engaged and non-engaged. The set of around 2,000 labeled images was used to train the ML model which will classify the given image into one of the categories. This system uses boosted decision tree regression which is a supervised learning method to predict the engagement level of un-labeled images (Soloviev, 2018).

The facial recognition systems are being currently used by educational institutions primarily for security reasons. However, it has a vast potential to assess the basic levels of affective domain skills. From a surveillance perspective, it is believed that it will constrain a healthy and unbiased environment which is very essential for an educational institution (Andrejevic & Selwyn, 2020). The current development of these techniques is also restricted to the lower levels of learning assessment i.e., attentiveness and responsiveness of learner.

5.4   PSYCHOMOTOR DOMAIN

The psychomotor domain is characterized by progressive levels of motor skills from perception to the mastery of a physical skill. Psychomotor competencies range from simple manual tasks like brushing teeth, to complicated tasks like operating an equipment or machinery. The complete spectrum of psychomotor competencies can be categorized into seven levels as shown in Figure 5.5. It is essential to acquire the competency at base level before moving to the next level. The most basic psychomotor

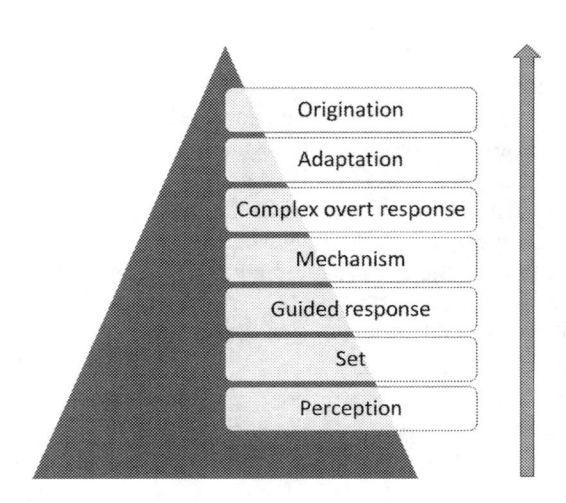

FIGURE 5.5   Hierarchy of the Psychomotor Domain Related to Learning.

competency is the perception or awareness of the surrounding environment (perception). After that, there is an individual's ability to respond physically, mentally and emotionally to a specific situation (set). The individuals acquire complex motor skills through series of stages; with initial emphasis on the imitation and trial-error (guided response). The individual gains a certain level of proficiency through practice (mechanism), and finally achieves accurate and highly coordinated expert-like performance (complex overt response). Therafter, at a certain level, individuals can extend the motor skills to fulfill the special requirements (adaptation) and lastly, individuals achieve the ability to create completely new patterns to justify new situations or to tackle specific problems (Simpson, 1966).

Innovations in the field of AI can contribute beyond human capabilities towards the development of tools and techniques which can accelerate the learning and assessment of the psychomotor skills. The advanced interactive technologies together with AI have a potential to develop learning and assessment systems for the activities like playing a musical instrument, performing surgical operations, drawing, dancing and playing sports (Santos, 2016). A virtual reality (VR) headset is a popular device mostly used in video gaming. However, it has very interesting applications in the field of education (Kleven et al., 2014). It is a head-mounted device which provides the feel of being in the digital world with the help of sound, images and motion tracking sensors. Another device, a leap motion controller is a commercially available portable device which is used to manipulate digital objects with hand motion. It makes use of infrared light along with optical sensors to recognize hand motion and gestures. Arifiani et al. proposed a technique for geometry validation using AI for 3D environments which involves the use of a VR headset and leap motion controller. An application was developed to evaluate psychomotor skills required for writing and drawing among elementary school students. The students were asked to redraw various geometric objects in a virtual environment by using a VR headset and leap motion controller. The objects drawn by the student were evaluated based on the closeness with the original object and a score was awarded accordingly. Though the system is very helpful for teaching and assessment of motor skills required for drawing objects at the elementary school level, the necessity of many devices and complicated installation processes keep it away from widespread use on the field (Arifiani et al., 2021).

Apart from leap motion controllers, the use of inertial sensors is reported in literature to capture human motion. Inertial sensors are the kind of transducers capable of detecting and measuring the slightest change in orientation, position and acceleration of an object. The compact size, low cost, high performance and precision makes it an excellent alternative to be used for capturing human motion data. In education, these inertial sensors are available as wearable devices and can be used to collect data about learners' body movements. Data about the body motion becomes a basis for the measurement of motor skills. In order to achieve the assessment of motor skills, it is essential to develop a model of human motion as a pattern recognition problem based on the data obtained from inertial sensors. The output of inertial sensor devices is a continuous signal in the time domain. These continuous signals are processed to extract the features which are used as input for the AI algorithms. The algorithms use these features for modeling human movements. Finally, learning performance for psychomotor skills can be assessed by validating the specific motion learning units of learner data with the data extracted from experts (Santos, 2019).

There is an increasing use of AI in the field of medical education (Imran & Jawaid, 2020). Performing a surgical procedure requires multiple psychomotor skills. Traditionally, trainee surgeons used to learn these skills under the supervision of expert surgeons with the help of animal or human objects. Also, the assessment of surgical performance of a trainee surgeon is very subjective and performed by expert surgeons. It is very crucial to acquire expertise in the surgical procedure before a surgeon could be allowed to handle it independently (Oropesa et al., 2011). The use of virtual reality (VR) simulators is gaining importance over time, and now trainee surgeons can practice surgical procedures with the help of VR simulators that mimic a surgical scene with human organs, creating a virtual scenario which is very close to reality. The VR simulator allows the trainee surgeon to learn the surgical procedures in a risk-free environment requiring minimum supervision from an expert. Various research-based and commercial VR simulators are available, such as MIST-simulator (Mentice Inc., Göteborg, Sweden) (Ahlberg et al., 2005), LAP Mentor (Simbionix USA Corp.) (Ayodeji et al., 2007), LapSim (Surgical Science Ltd., Gothenburg, Sweden) (van Dongen et al., 2007) and SIMENDO (DelltaTech, Delft, The Netherlands) (Verdaasdonk et al., 2007). The training component of these simulators is further extended for the assessment of trainee surgeons using several ML techniques. These techniques are capable of assessing surgical competencies of surgeon mainly involving two phases. In the first phase human motion data is collected from the VR simulator while performing a specific surgical procedure for the expert and non-expert surgeons. The ML model is trained using this data to establish the competency model. In the next phase, data extracted for the unknown learner is compared with the model created in the first phase. Based on this comparison, new surgeons can be classified in various competency levels such as novice, intermediate or expert (Hajshirmohammadi & Payandeh, 2007; Megali et al., 2006; Westwood, 2005).

Alonso-Silverio et al. proposed an artificial neural network (ANN)-based hybrid box trainer tool for the assessment of basic laparoscopic surgery skills. The laparoscopic surgery is a modern surgical procedure wherein the surgeon can operate on the organs inside the abdomen or pelvis without making large cuts on the skin. It is a type of minimally invasive surgery which makes use of small diameter narrow tubes which are inserted in the body. These tubes allow torches, small cameras and surgical instruments to access and operate the internal organs. The surgeon manipulates the surgical instruments with the help of internal view of body displayed on screen which is captured with the help of camera and light source. Traditionally, surgeons were trained for laparoscopic surgery with the help of a physical laparoscopic box trainer. The modern alternative to the physical model is the virtual reality laparoscopic trainer systems. These simulation systems allow trainers to perform the surgical procedure in a completely virtual environment. However, there are certain advantages of using a physical model for training over a virtual simulator as it provides a more realistic environment with real laparoscopic surgery instruments. The hybrid laparoscopic trainer is the best alternative which combines the benefits of both physical and simulated training. One of the most important advantages of a hybrid trainer is that the trainees can be objectively evaluated for their psychomotor skills with the help of intelligent software systems. This hybrid trainer system was designed to assess two basic surgical tasks: transferring and pattern cutting. Transferring involves the skill of a trainee surgeon to move the virtual tip of a surgical instrument from an original point to a target point in a three-dimensional space, whereas pattern cutting involves hand and eye coordination of a trainee surgeon in tracing an ideal spiral with the help of the virtual tip of a surgical instrument. The data for these two tasks was collected for 20 medical professionals that belongs to two classes—experts and trainee surgeons. The binary images of the spiral traces were recorded for the surgical task of pattern cutting and the same images were used as a training dataset for the ANN model. The aim of the ANN model is to predict the level of an individual's psychomotor skills for the pattern cutting tasks based on binary images of traces. The ANN model is capable of classifying whether the given trace from a binary image belongs to an expert surgeon or non-expert surgeon with an accuracy of 93.94% (Alonso-Silverio et al., 2018). Such systems have potential to assess the motor skills for complex tasks like surgical procedures. The pilot studies reported in the literature are demonstrating simple psychomotor skills like tracing a pattern with the help of virtual tools. It is very essential to extend these assessment models for more complicated tasks.

5.5 CHALLENGES OF AI IN LEARNING ASSESSMENT

The incorporation of AI in learning assessment can offer unintended and unanticipated consequences. These consequences may not surface immediately as it may take substantial time to notice them (Pringle et al., 2016). An AI-based student performance prediction system is considered to be one of the most promising applications of AI in the educational system. This type of system offers benefits in terms of identifying the students who are underperforming and are at the risk of being dropped out. However, there are associated risks in predicting such outcomes related to students' performance. The algorithms and training datasets employed to build such systems may carry forward existing biases such as human biases related to gender and ethnicity. Also, such predictions will affect the individual's learning path and future choices which are dictated by a non-human entity (Berendt et al., 2020). AI-driven systems collect a large amount of data as a part of their functioning, and hence, there are serious concerns related to the collection, analysis and use of learners' personal data by human-rights advocates (Lupton & Williamson, 2017). Though AI-enabled assessment is a fascinating idea, the cost, time, and expertise required to design and implement an intelligent assessment system can be a major challenge, as it involves computer programmers, domain experts and educational theorists (Beck et al., 1996).

5.6 CONCLUSION

The educational system has well adopted automation in the assessment and evaluation of students' learning. However, it is limited to computerized objective tests in the form of multiple choice, true/false answers, multiple-response and matching questions. Again, this automation largely addresses only the basic level of cognitive skills. To achieve automated assessment across all the domains of learning, it is essential to explore the field of AI. In this chapter, we have taken a review of automated intelligent assessment (AIA) system in the three domains of learning—cognitive, affective and psychomotor.

The cognitive domain is mainly concerned with students' ability in the comprehension of knowledge and various intellectual skills. The Bloom's taxonomy is widely used as a reference point for the assessment and evaluation of these cognitive skills. Various AI techniques like ML and NLP are gaining importance for the development of assessment tools for various cognitive skills. Auto grading technology for short answers and essays are capable of automating the assessment of constructed responses by the students. These techniques have raised the level of automation from objective to subjective types of questions. There is still scope for the development of automated intelligent assessment tools for higher order thinking skills.

The affective domain of learning is concerned with how emotions, feelings and attitude affect student learning. On the contrary, psychomotor domain deals with bodily movement, coordination and use of the motor-ability. Though the assessment of students' affective and psychomotor competencies is very crucial, often it is neglected due to its intangible nature. The use of IP, voice recognition, NLP, virtual reality, inertial sensors and physiological signals along with various AI techniques to measure the affective and psychomotor state of students, has provided us an opportunity to get quantitative data for assessment. When we consider the complete assessment spectrum for all three learning domains, AI techniques have been implemented for the assessment of basic levels of skills. However, there is still room to develop the tools which will address the assessment of higher level of skills in all three learning domains.

REFERENCES

Ahlberg, G., Kruuna, O., Leijonmarck, C.-E., Ovaska, J., Rosseland, A., Sandbu, R., Strömberg, C., & Arvidsson, D. (2005). Is the learning curve for laparoscopic fundoplication determined by the teacher or the pupil? *The American Journal of Surgery*, *189*(2), 184–189. https://doi.org/10.1016/j.amjsurg.2004.06.043.

Alonso-Silverio, G. A., Pérez-Escamirosa, F., Bruno-Sanchez, R., Ortiz-Simon, J. L., Muñoz-Guerrero, R., Minor-Martinez, A., & Alarcón-Paredes, A. (2018). Development of a laparoscopic box trainer based on open source hardware and Artificial Intelligence for objective assessment of surgical psychomotor skills. *Surgical Innovation*, *25*(4), 380–388. https://doi.org/10.1177/1553350618777045.

Anderson, L. W., Krathwohl, D. R., & Bloom, B. S. (2001). *A taxonomy for learning, teaching, and assessing: A revision of Bloom's taxonomy of educational objectives* (Complete ed.). New York: Longman.

Andrejevic, M., & Selwyn, N. (2020). Facial recognition technology in schools: Critical questions and concerns. *Learning, Media and Technology*, *45*(2), 115–128. https://doi.org/10.1080/17439884.2020.1686014.

Arifiani, S., Yuhana, U. L., Hariadi, R. R., & Andrianto, F. P. (2021). *The implementation of Artificial Intelligence in geometry validation for psychomotor aspect assessment*. International Conference on Educational Assessment and Policy (ICEAP 2020), pp. 200–204.

Arora, Y., Singhal, A., & Bansal, A. (2014). Prediction & warning: A method to improve student's performance. *ACM SIGSOFT Software Engineering Notes*, *39*(1), 1–5. https://doi.org/10.1145/2557833.2557842.

Ayodeji, I. D., Schijven, M., Jakimowicz, J., & Greve, J. W. (2007). Face validation of the Simbionix LAP Mentor virtual reality training module and its applicability in the surgical curriculum. *Surgical Endoscopy, 21*(9), 1641–1649. https://doi.org/10.1007/s00464-007-9219-7.

Bălan, O., Moise, G., Moldoveanu, A., Leordeanu, M., & Moldoveanu, F. (2019). Fear level classification based on emotional dimensions and machine learning techniques. *Sensors, 19*(7), 1738. https://doi.org/10.3390/s19071738.

Beck, J., Stern, M., & Haugsjaa, E. (1996). Applications of AI in education. *XRDS: Crossroads, The ACM Magazine for Students, 3*(1), 11–15. https://doi.org/10.1145/332148.332153.

Berendt, B., Littlejohn, A., & Blakemore, M. (2020). AI in education: Learner choice and fundamental rights. *Learning, Media and Technology, 45*(3), 312–324. https://doi.org/10.1080/17439884.2020.1786399.

Bloom, B. S. (1956). *Taxonomy of educational objectives, handbook 1: Cognitive domain* (2nd ed.). New York: Addison-Wesley Longman Ltd.

Braiki, B. A., Harous, S., Zaki, N., & Alnajjar, F. (2020). Artificial Intelligence in education and assessment methods. *Bulletin of Electrical Engineering and Informatics, 9*(5), 1998–2007. https://doi.org/10.11591/eei.v9i5.1984.

Cicchinelli, A., Veas, E., Pardo, A., Pammer-Schindler, V., Fessl, A., Barreiros, C., & Lindstädt, S. (2018). Finding traces of self-regulated learning in activity streams. *Proceedings of the 8th International Conference on Learning Analytics and Knowledge*, 191–200. https://doi.org/10.1145/3170358.3170381.

Costa, E. B., Fonseca, B., Santana, M. A., de Araújo, F. F., & Rego, J. (2017). Evaluating the effectiveness of educational data mining techniques for early prediction of students' academic failure in introductory programming courses. *Computers in Human Behavior, 73*, 247–256. https://doi.org/10.1016/j.chb.2017.01.047.

Dzikovska, M. O., Nielsen, R. D., Brew, C., Leacock, C., Giampiccolo, D., Bentivogli, L., Clark, P., Dagan, I., & Dang, H. T. (2013). *Semeval-2013 task 7: The joint student response analysis and 8th recognizing textual entailment challenge*. Denton, TX: North Texas State University Denton.

Epstein, M. L., Lazarus, A. D., Calvano, T. B., Matthews, K. A., Hendel, R. A., Epstein, B. B., & Brosvic, G. M. (2002). Immediate feedback assessment technique promotes learning and corrects inaccurate first responses. *The Psychological Record, 52*(2), 187–201. https://doi.org/10.1007/BF03395423.

Gardner, J., O'Leary, M., & Yuan, L. (2021). Artificial Intelligence in educational assessment: 'Breakthrough? Or buncombe and ballyhoo?' *Journal of Computer Assisted Learning, 37*(5), 1207–1216.

Geigle, C., Zhai, C., & Ferguson, D. C. (2016). An exploration of automated grading of complex assignments. *Proceedings of the Third (2016) ACM Conference on Learning @ Scale*, 351–360. https://doi.org/10.1145/2876034.2876049.

Hajshirmohammadi, I., & Payandeh, S. (2007). Fuzzy set theory for performance evaluation in a surgical simulator. *Presence: Teleoperators and Virtual Environments, 16*(6), 603–622. https://doi.org/10.1162/pres.16.6.603.

Haudek, K. C., Prevost, L. B., Moscarella, R. A., Merrill, J., & Urban-Lurain, M. (2012). What are they thinking? Automated analysis of student writing about acid—base chemistry in introductory biology. *CBE—Life Sciences Education, 11*(3), 283–293. https://doi.org/10.1187/cbe.11-08-0084.

Holt, B. J., & Hannon, J. C. (2006). Teaching-learning in the affective domain. *Strategies, 20*(1), 11–13. https://doi.org/10.1080/08924562.2006.10590695.

Imran, N., & Jawaid, M. (2020). Artificial Intelligence in medical education: Are we ready for it? *Pakistan Journal of Medical Sciences, 36*(5), 857–859. https://doi.org/10.12669/pjms.36.5.3042.

Jain, A. K., Duin, R. P. W., & Mao, J. (2000). Statistical pattern recognition: A review. *IEEE Transactions on Pattern Analysis and Machine Intelligence*, *22*(1), 4–37. https://doi.org/10.1109/34.824819.

Kleven, N. F., Prasolova-Førland, E., Fominykh, M., Hansen, A., Rasmussen, G., Sagberg, L. M., & Lindseth, F. (2014). *Training nurses and educating the public using a virtual operating room with oculus rift*. 2014 International Conference on Virtual Systems Multimedia (VSMM), pp. 206–213. https://doi.org/10.1109/VSMM.2014.7136687.

Kraftwohl, D. R., Bloom, B. S., & Masia, B. B. (1967). *Taxonomy of educational objectives, the classification of educational goals: Handbook II: Affective domain*. New York: David McKay Company.

Leacock, C., & Chodorow, M. (2003). C-rater: Automated scoring of short-answer questions. *Computers and the Humanities*, *37*(4), 389–405. https://doi.org/10.1023/A:1025779619903.

Leonard Liaw, H., Chiu, M.-H., & Chou, C.-C. (2014). Using facial recognition technology in the exploration of student responses to conceptual conflict phenomenon. *Chemistry Education Research and Practice*, *15*(4), 824–834. https://doi.org/10.1039/C4RP00103F.

Lin, H.-C. K., Wang, C.-H., Chao, C.-J., & Chien, M.-K. (2012). Employing textual and facial emotion recognition to design an affective tutoring system. *Turkish Online Journal of Educational Technology—TOJET*, *11*(4), 418–426.

Lupton, D., & Williamson, B. (2017). The datafied child: The dataveillance of children and implications for their rights. *New Media & Society*, *19*(5), 780–794. https://doi.org/10.1177/1461444816686328.

Manhães, L. M. B., da Cruz, S. M. S., & Zimbrão, G. (2014). *WAVE: An architecture for predicting dropout in undergraduate courses using EDM*. Proceedings of the 29th Annual ACM Symposium on Applied Computing, pp. 243–247. https://doi.org/10.1145/2554850.2555135.

Megali, G., Sinigaglia, S., Tonet, O., & Dario, P. (2006). Modelling and evaluation of surgical performance using hidden Markov models. *IEEE Transactions on Biomedical Engineering*, *53*(10), 1911–1919. https://doi.org/10.1109/TBME.2006.881784.

Mitchell, T., Russell, T., Broomhead, P., & Aldridge, N. (2002). *Towards robust computerised marking of free-text responses*. Proceedings of the 6th CAA Conference, Loughborough: Loughborough University.

Moharreri, K., Ha, M., & Nehm, R. H. (2014). EvoGrader: An online formative assessment tool for automatically evaluating written evolutionary explanations. *Evolution: Education and Outreach*, *7*(1), 15. https://doi.org/10.1186/s12052-014-0015-2.

Oropesa, I., Sánchez-González, P., Lamata, P., Chmarra, M. K., Pagador, J. B., Sánchez-Margallo, J. A., Sánchez-Margallo, F. M., & Gómez, E. J. (2011). Methods and tools for objective assessment of psychomotor skills in laparoscopic surgery. *Journal of Surgical Research*, *171*(1), e81–e95. https://doi.org/10.1016/j.jss.2011.06.034.

Page, E. B. (1966). The imminence of . . . Grading essays by computer. *The Phi Delta Kappan*, *47*(5), 238–243.

Pena-Shaff, J. B., & Nicholls, C. (2004). Analyzing student interactions and meaning construction in computer bulletin board discussions. *Computers & Education*, *42*(3), 243–265. https://doi.org/10.1016/j.compedu.2003.08.003.

Pringle, R., Michael, K., & Michael, M. G. (2016). Unintended consequences of living with AI: The paradox of technological potential? Part II [Guest Editorial]. *IEEE Technology and Society Magazine*, *35*(4), 17–21.

Ramachandran, L., Gehringer, E. F., & Yadav, R. K. (2017). Automated assessment of the quality of peer reviews using natural language processing techniques. *International Journal of Artificial Intelligence in Education*, *27*(3), 534–581. https://doi.org/10.1007/s40593-016-0132-x.

Robinson, R. (2001). Calibrated peer review™ an application to increase student reading & writing skills. *The American Biology Teacher*, *63*(7), 474–480. https://doi.org/10.2307/4451167.

Santos, O. C. (2016). Training the body: The potential of AIED to support personalized motor skills learning. *International Journal of Artificial Intelligence in Education*, *26*(2), 730–755.

Santos, O. C. (2019). Artificial Intelligence in psychomotor learning: Modeling human motion from inertial sensor data. *International Journal on Artificial Intelligence Tools*, *28*(4), 1940006. https://doi.org/10.1142/S0218213019400062.

Simpson, E. J. (1966). *The classification of educational objectives, psychomotor domain.* Available at: https://eric.ed.gov/?id=ED010368.

Soloviev, V. (2018). Machine learning approach for student engagement automatic recognition from facial expressions. *Scientific Publications of the State University of Novi Pazar Series A: Applied Mathematics, Informatics and Mechanics*, *10*(2), 79–86. https://doi.org/10.5937/SPSUNP1802079S.

van Dongen, K. W., Tournoij, E., van der Zee, D. C., Schijven, M. P., & Broeders, I. A. M. J. (2007). Construct validity of the LapSim: Can the LapSim virtual reality simulator distinguish between novices and experts? *Surgical Endoscopy*, *21*(8), 1413–1417. https://doi.org/10.1007/s00464-006-9188-2.

Verdaasdonk, E. G. G., Stassen, L. P. S., Schijven, M. P., & Dankelman, J. (2007). Construct validity and assessment of the learning curve for the SIMENDO endoscopic simulator. *Surgical Endoscopy*, *21*(8), 1406–1412. https://doi.org/10.1007/s00464-006-9177-5.

Vittorini, P., Menini, S., & Tonelli, S. (2020). An AI-based system for formative and summative assessment in data science courses. *International Journal of Artificial Intelligence in Education*. https://doi.org/10.1007/s40593-020-00230-2.

Vogler, K. E. (2005). Improve your verbal questioning. *The Clearing House: A Journal of Educational Strategies, Issues and Ideas*, *79*(2), 98–103. https://doi.org/10.3200/TCHS.79.2.98-104.

Westwood, J. D. (2005). *Medicine meets virtual reality 13: The magical next becomes the medical now.* Amsterdam, NY: IOS Press.

Yahya, A. A., Osman, A., Taleb, A., & Alattab, A. A. (2013). Analyzing the cognitive level of classroom questions using machine learning techniques. *Procedia—Social and Behavioral Sciences*, *97*, 587–595. https://doi.org/10.1016/j.sbspro.2013.10.277.

Yang, C.-H. (2012). Fuzzy fusion for attending and responding assessment system of affective teaching goals in distance learning. *Expert Systems with Applications*, *39*(3), 2501–2508. https://doi.org/10.1016/j.eswa.2011.08.102.

Zhai, X., Chu, X., Chai, C. S., Jong, M. S. Y., Istenic, A., Spector, M., Liu, J.-B., Yuan, J., & Li, Y. (2021). A review of Artificial Intelligence (AI) in education from 2010 to 2020. *Complexity*, *2021*, e8812542. https://doi.org/10.1155/2021/8812542.

Zhai, X., Fang, Q., Dong, Y., Wei, Z., Yuan, J., Cacciolatti, L., & Yang, Y. (2018). The effects of biofeedback-based stimulated recall on self-regulated online learning: A gender and cognitive taxonomy perspective. *Journal of Computer Assisted Learning*, *34*(6), 775–786. https://doi.org/10.1111/jcal.12284.

6 AI-Based Predictive Models for Adaptive Learning Systems

Prashant Gupta, Trishul Kulkarni and Bhagwan Toksha

CONTENTS

6.1 INTRODUCTION

Artificial Intelligence (AI), first coined by computer scientist John McCarthy, can be defined as a computer system with the ability to conduct various human tasks such as reasoning, listening, problem-solving, communicating, and more, at a cognitive level (Baker et al., 2019). It can be considered as science fiction jargon which has become an integral part of our lives via its applications in healthcare, logistics, finance, banking, and so on. It has revolutionized the way we do work in today's world and the education sector has also seen a fair share of its use. AI has come out of just the ideal "thought and planned laboratory situations" to address more complex real-life learning ones. The market in the domain of educational technology has enabled industries to develop individual adaptive learning systems that help in customizing the personalized learning experience. These types of systems aid in managing the classroom environment, marking and evaluating, along with catering to the second-language issue. In a period between 2008 and 2017, the angel investment into AI-based education globally was around $1,047 billion (Mou, 2019). However, issues such as integration of AI-based education systems within macro- and micro educational ecosystems, clarity over the stakeholder role in such a learning environment, along with business versus consumer development, have presented challenges. Some of the questions that arise are:

- Are these issues discussed in the AI-based education literature?
- How is the educational paradigm to progressively evolve with role changes in AI technology?
- Which direction is the research in AI-based education technology moving towards?

Although classrooms have already been majorly digitized, the physical grind of teachers and learners involved in the preparation of course content, providing and appearing for assignments, grading them and providing feedback to learners and their guardians on progress, has remained the same. This has led to an obstacle in the learning activities which is not personalized to their needs, context, and abilities. It is important as in a classroom with any number of learners, they vary from learner to learner, and a "one size fits all" learning system may not be promising. Late Sir Ken Robinson, a New York Times best-selling author, who led national and international projects on creative and cultural education quoted in one of his TED talks, "Why is there this assumption that we should educate children simply according to how old they are? It's almost as if the most important thing that children have in common is their date of manufacture".

The historical "hand wrought" challenges can be addressed by recent advances in AI and machine learning (ML) by bringing in a large amount of data including learner demographics, teacher demographics, performance, human resources, registration and admission information for learner background, and more. Machine learning works well in the case of prediction and estimation, which shall be discussed in further sections. This chapter will encompass predictive modeling basics, types of learning and a detailed account of intelligent tutoring systems, their design and challenges, and will comprise predictive models and elaborations based on some commercial models as illustrations.

6.2 MODERN-DAY EDUCATION AND THE ROLE OF AI IN TEACHING-LEARNING

From a technical forefront, the world is moving at a rapid pace, and academics must keep up with the pace. Some of the changes that are happening/bound to happen in the academic world due to technological developments in education are illustrated in Figure 6.1 and explained to follow:

- Learning may no longer be defined by time and place. It may happen at the liberty of the learner unless he/she insists it be traditional (classroom-oriented) in nature.

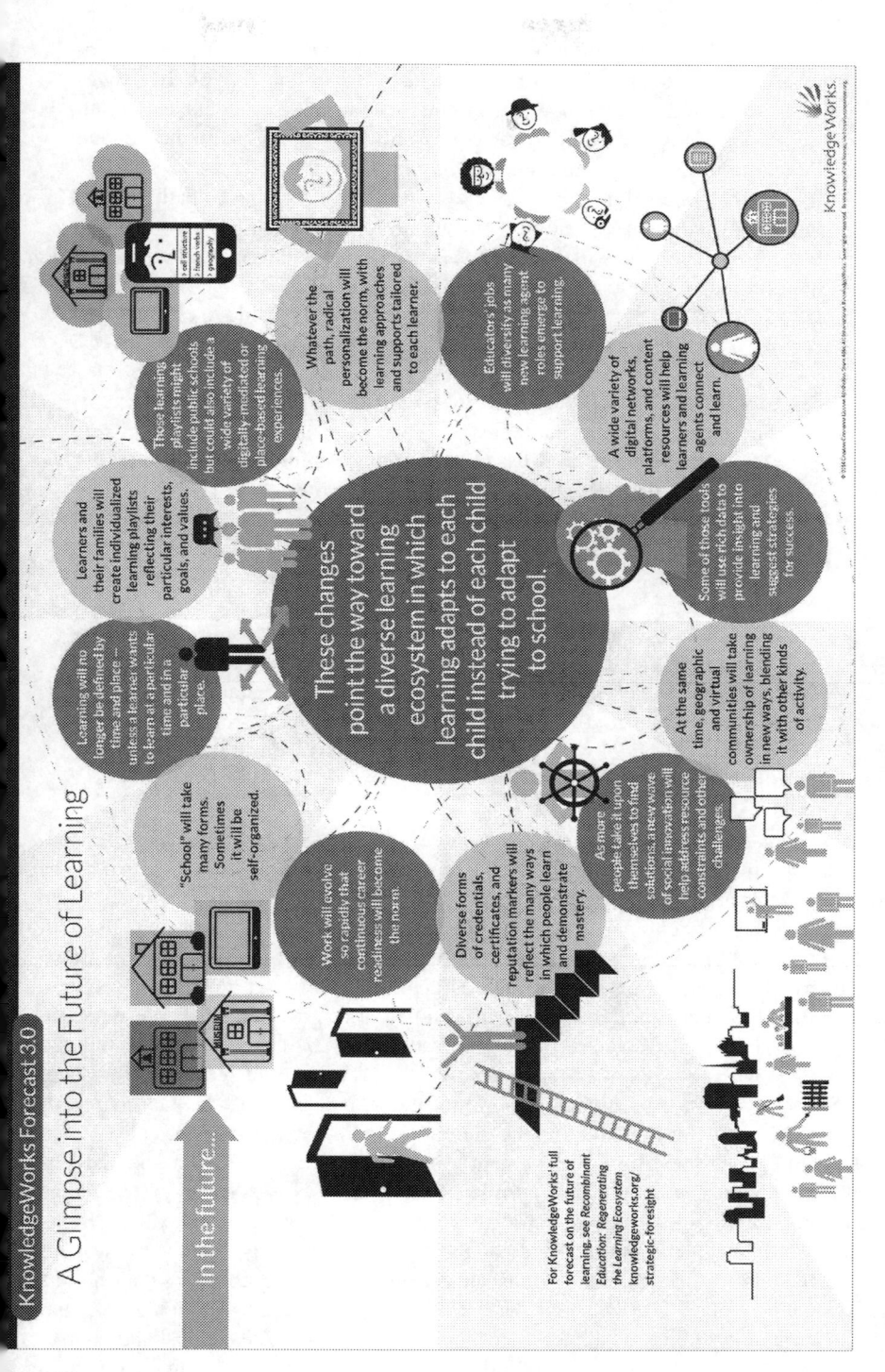

FIGURE 6.1 A Glimpse into the Future of Learning (Knowledge Works, 2013).

- Learning will be based on individual learning interests, goals, and values, with an option to create an individual course content playlist on their own.
- These so-called learning playlists might be inclusive of content used in public schools. However, a large variety of online digital educational content, including place-based experiences will be part of it.
- Radical personalization will prevail irrespective of the chosen path, with an individualized approach along with support towards each learner.
- Educationalists' role will diversify as new learning agents emerge to support the process of learning.
- Learners and the agents involved in the process may be able to connect over various digital platforms, networks and content-based resources to increase the effectiveness of the learning process.
- The tools that will be employed will use rich data to look into learning activities and identify success strategies.
- The involvement in position-based and virtual communities will be pivotal and synchronized at the same time for introducing new ways of learning and blending it with activity-based methods.
- A new tide of society-oriented innovations from an increasing number of entrepreneurs will aid in addressing limitations regarding resources, amongst others.
- The current certificate system will get diversified with various methodologies of learning and demonstrating mastery in various trades.
- The readiness to serve in one's career will take center stage as work will keep evolving on a continual basis.
- The concept of school will be learner-dependent as well. It may sometimes be self-organized.

These changes have directed academic education towards a diverse learning ecosystem in which learning/course content must adapt to learner needs, which is the opposite of what is going on. In order for this adaptation, the use of computer-based systems has a big role to play. Although computers have found themselves in the education sector, AI-based systems are still in the nascent stages in terms of fully-grown commercial use. However, sectors such as healthcare, agriculture, sports, services, banking and finance, retail, and so on offer a gamut of products incorporating technologies based on AI (Benke & Benke, 2018; Boobier, 2020; Cao, 2020; Deshmukh, 2018; M.-H. Huang & Rust, 2018; Jarrahi, 2018; Leyva-Vázquez et al., 2018). Baker and Smith categorized AI applications in education on the basis of orientation towards the learner, instructor or institutional system. A learner-oriented AI is based on a personalized/adaptive learning management system (LMS), an instructor-oriented system that aids in setting up tasks and related assessments. Institutional system-based assessment can help with administrative tasks and procedures in education (Baker et al., 2019). With the development of individualized intelligent tutoring systems (Khazanchi & Khazanchi, 2021; Mousavinasab et al., 2021), teacher bots (Georgescu, 2018; Sandu & Gide, 2019), support platforms (Ciolacu et al., 2018), and more, AI has demonstrated its ability to help learners by offering personalized support and coursework on the basis of identification of individual knowledge gaps. This in turn reduces the instructor workload from laborious daily tasks which enables faster individual learner responses from their side. Also, it provides insightful help to policymakers, for instance, regarding attrition patterns across various streams or colleges and enrolment behaviour (Wang, 2021). Certain subsets of AI—ML, Neural Networks (NN), and Deep Learning (DL)—have found their way into the predictive modelling arena, which is important in order to offer an individualized learning approach in the educational sector (Sood & Saini, 2021; Tsai et al., 2020). Some of the benefits that they have brought to the education community are:

- Learning Analytics, which create statistical models of learner knowledge to provide personalized, computer-aided feedback on the learning progress of learners and their instructors.
- Content Analytics that enhance and organize content items such as quizzes, course assessments, textbook sections, course-based educational videos, and so on.
- Programming algorithms that seek optimal and tailored instructional policies that help learners learn more efficiently.
- Grading systems that rate and evaluate learner responses to large-scale computer assessments and assignments either automatically or through peer evaluation.
- Cognitive psychology, where data mining is increasingly becoming an important tool for validating cognitive science theories and facilitating the development of new ones to improve the process of learning and associated retention of knowledge.
- Aid for active learning and experimental design, which is unique to each individual learner on an adaptive basis, thereby enhancing learning efficiency on the basis of individual selection of assessment and learning resources.

6.3 PREDICTIVE MODELLING AND THEIR ALGORITHMS

Predictive modelling is a well-researched and deep-rooted area in AI education. It can be considered as predicting outcomes with a host of techniques employed by creating inferences regarding uncertain events that can happen in the future or data modelling. In the education sector, an administrator may be interested in predicting the measurement for learning (which can be acquiring of skills, academic success regarding the learners), teaching (which can be the effect of a specific teaching faculty or an instructional strategy on an individual learner) and other matrices such as retention or course registration predictions. An array of various products such as Ellucian, Blackboard, and Starfish Retention Solutions are available which incorporate predictive analytics in the LMS. Various other companies such as Blue Canary and Civitas Learning consult for customizing higher education solutions. Educational institutions can utilize these models to provide real-time learning needs. The intent is to create a setup that would be accurately able to designate uninterrupted learner outcomes. For example, it can be employed to understand the likelihood of a given individual to complete their educational degree by applying this model to all learners on an individual basis and determining when they might complete their academic degrees with an assumption that no intervention strategy is applied.

Predictive modelling is different from explanatory modelling as it is based on the assumption that the training instance data, also referred to as a set of known data, can be utilized to predict the class or value of new data based on the variables observed. It is an in situ activity that is planned to make systems responsive to changes in the underlying data. Brooks et al. described a learner success system based on a predictive modelling approach using data collected from clickstream data; in this case, historical log files (Brooks et al., 2015).

The process of predictive analytics can be organized into three stages: analysis, monitoring and prediction. The analysis phase is involved in the collection of raw data (pre-processed) and its transformation by analysis to make a new model. The monitoring phase involves the observation of analyzed data and the creation of a learning methodology using it. The third and last phase is prediction which uses the learning methodology and ML algorithms to predict the required events and report them to the users. To understand the actual workflow for a predictive modelling process, Gladshiya and Sharmila have represented it by a seven stage process consisting of defining the project, collection of data, cleaning, deep analysis, use of statistical regression methods, employing machine learning techniques, construction of a model, its deployment and simultaneous monitoring (Gladshiya & Sharmila, 2019). A descriptive schematic of predictive modelling workflow can be seen in Figure 6.2.

Predictive analytics largely overlap with ML as they require an ML algorithm (Rastrollo-Guerrero et al., 2020). ML as quoted earlier is an area of study that enables the computer algorithms to learn without the need to programme them explicitly. It has evolved from pattern identification study and explores the potential of algorithms to learn from data and make predictions. Also, in the process, they become more intelligent and can overcome instructions to take data-driven accurate decisions. There are two general categories of prediction models as regression and pattern classification under supervised learning, which comprises algorithms. These algorithms are involved in data mining and statistical analysis, which in turn help in the identification of certain patterns and trends in the data. Supervised learning problems are where a data set is given and the researcher has a fair amount of idea as to what the output should be like with knowledge of the relationship between the input and its respective output.

a. *Regression Models:* are designed for predicting numbers. The attempt in a regression type problem is to predict results with a continuous output. Regression analytics investigate the relationship between a dependent and independent variable, target and predictor, in this case, respectively, and can be employed for forecasting, time series modelling and determining cause and effect relationships between the variables involved. Some of the educational problems that these models can touch upon are how parents and siblings affect the institutional outcomes of a learner, how climate and socioeconomic composition of school and climate, respectively, impact average achievements of the educational institution, and so on. The most commonly used regressions are linear, logistic, polynomial, stepwise, ridge,

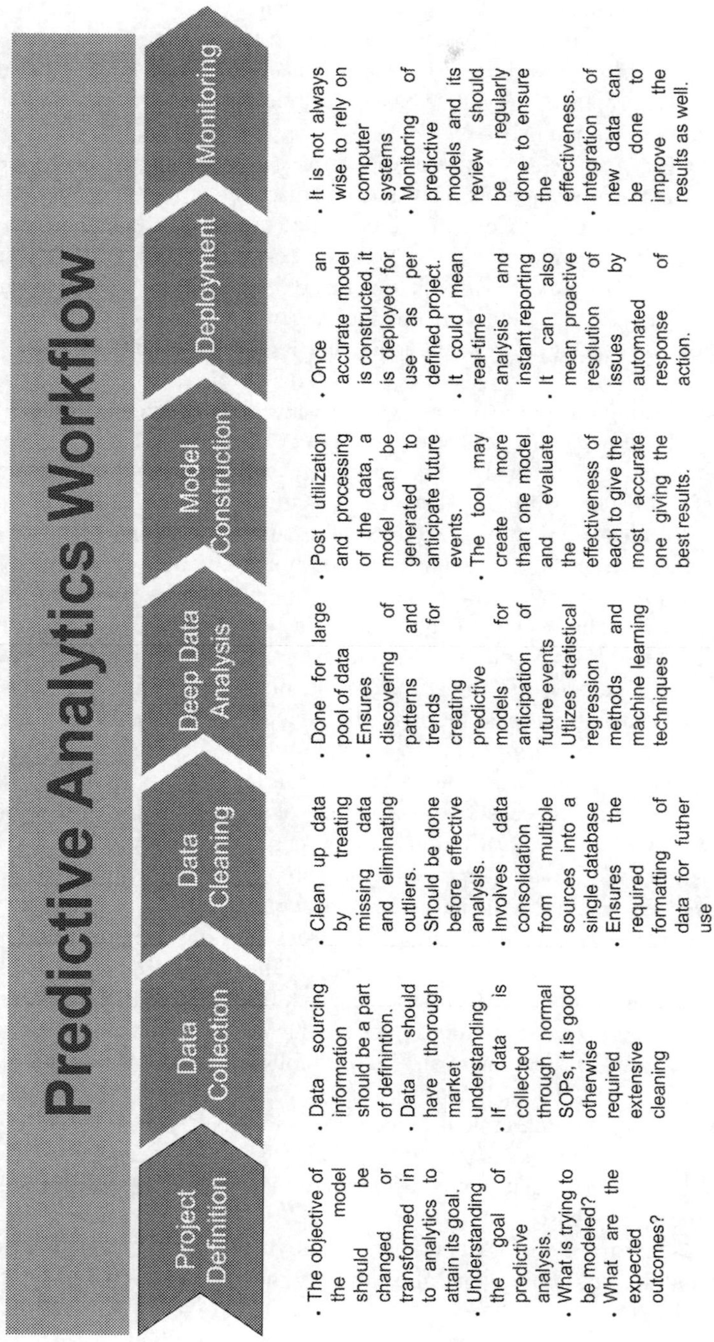

Predictive Analytics Workflow

Project Definition	Data Collection	Data Cleaning	Deep Data Analysis	Model Construction	Deployment	Monitoring
• The objective of the model should be changed or transformed in to analytics to attain its goal. • Understanding the goal of predictive analysis. • What is trying to be modeled? • What are the expected outcomes?	• Data sourcing information should be a part of definition. • Data should have thorough market understanding • If data is collected through normal SOPs, it is good otherwise required extensive cleaning	• Clean up data by treating missing data and eliminating outliers. • Should be done before effective analysis. • Involves data consolidation from multiple sources into a single database • Ensures the required formatting of data for further use	• Done for large pool of data • Ensures discovering of patterns and trends for creating predictive models for anticipation of future events • Utilzes statistical regression methods and machine learning techniques	• Post utilization and processing of the data, a model can be generated to anticipate future events. • The tool may create more than one model and evaluate the effectiveness of each to give the most accurate one giving the best results.	• Once an accurate model is constructed, it is deployed for use as per defined project. • It could mean real-time analysis and instant reporting • It can also mean proactive resolution of issues by automated response of action.	• It is not always wise to rely on computer systems • Monitoring of predictive models and its review should be regularly done to ensure the effectiveness. • Integration of new data can be done to improve the results as well.

FIGURE 6.2 The Workflow of Predictive Analytics for Predictive Modelling.

lasso, elasticnet and more. Huang and Fang attempted to predict learner academic performance in an engineering dynamics course (S. Huang & Fang, 2010). Alshanqiti and Nouman optimized the prediction accuracy of learner academic performance using a hybrid regression model and compared its performance with single baseline models such as linear regression (Alshanqiti & Namoun, 2020).

b. *Pattern Classification Models:* are responsible for predicting class membership with an attempt at predicting results in a discrete output—mapping input variables into discrete categories. A classification model will attempt to draw a conclusion from the values observed. Some of the educational issues that it can touch upon include college risk of getting failed in the first year of engineering education (Garg, 2021), factors affecting placements, productivity and learning goal orientation of engineering learners (Gray et al., 2014), and so on.

Unsupervised learning, in contrast, deals with non-labelled occurrences and the classifications must be drawn from an unstructured pool of data without human interventions. Thus, it employs a clustering technique for grouping the non-labelled samples on the basis of any pattern/similarity/distance measures. Chien et al. developed a prediction model for measurement of the degree of creativity in an engineering course based on discussion records of learners while learning (Chien et al., 2020). A subtype of unsupervised learning, reinforcement learning involves learning from a series of actions by maximizing a reward function. This function can be maximized by either rewarding good actions and vice-versa for bad ones. Shawaky and Badawi presented an approach capable to adapt to the most influential parameters dependent upon learner and learning settings, such as individual and collaborative learning with the use of reinforcement learning. This approach builds an intelligent environment to offer learning materials suitable to the individual learner. Also, it provides a methodology for comprehending learner state and technology acceptance, both changing with time (Shawky & Badawi, 2018). Liu et al. proposed recommendations for smart educational classrooms, which comprise the measurement of quiz scores, heartbeats, blinks and facial expressions of the individual learner to design the learning activities and apply reinforcement learning to suggest efficient learning activities for learners based on their current state of learning and acquired knowledge (Liu et al., 2018).

Some of the algorithms based on supervised, unsupervised and reinforcement learning that can be used for predictive modelling are listed (Asiah et al., 2019; Brooks & Thompson, 2017; Hasan et al., 2020) to follow:

1. Linear and Logistic Regression: Linear regression algorithms predict a continuous output (numeric) from a linear combination of attributes whereas logistic regression predicts the odds of two or more outcomes, thereby helping with category-based predictions.
2. Nearest Neighbor: These are classifiers employing only the nearest labelled data points in the training dataset to identify the right predicted labels for new data.
3. Decision Trees: These are the simplest and easiest methods for comprehending both large and small data. They are employed for multiple variable

analysis. They are produced of algorithms that identify different ways of data bifurcation into branch-like segments resembling a tree. They partition data into subsets based on the input variable categories to aid in understanding someone's decision path.

4. Neural Networks: These are biologically inspired artificial networks patterned similar to the function of neurosin, and are a part of deep learning methodology. They are employed for solving complex pattern recognition problems and are capable to be efficient in the analysis of large data sets. They can handle data with nonlinear relationships and function well with uncertain variables.

5. Time series algorithms: These are utilized for a sequential plot of data and can forecast continuous values over time.

6. Outlier detection algorithms: These focus on the detection of anomalies, identification of items, events or observations that are not conforming to an expected pattern or standard within the data.

7. Ensemble Models: These use multiple ML algorithms to improve the efficiency of prediction performance over that of what can be obtained from a single algorithm. Two prominent techniques are bootstrap aggregating, which involve the building of several prediction models from random sub-samples of the dataset, and boosting, wherein the design of successive predictive models is done to cover for the misclassifications of prior models. One such model is Random Forest, which is a classification/regression-based ML algorithm. It is flexible, easy to use and produces a great result most of the time even without tuning of the hyper-parameter. It is also one of the most widely used algorithms, due to its simplicity and diversity.

8. Naïve Bayes: This is a classifier that allows to predict a category/class on the basis of a given feature set with the use of probability. They assume the statistical independence of each attribute.

9. Bayesian Network: They feature a manually constructed graphical model reflecting the probabilistic dependence relationship between variables.

10. Support Vector Machine: This algorithm is competent in delivering high accuracy when compared to other algorithms of data classification type. It also has a good generalization and is faster than other methods.

There are many more algorithms used, which come under the larger blanket of ML, but the common ones used mostly in predictive modelling are discussed here.

6.4 MACHINE LEARNING IN PERSONALIZED LEARNING

ML has many applications in the field of education such as adaptive learning, increase in efficiency, learning analytics, predictive analytics, grading assignments accurately and personalized learning. One of the best usages of ML is personalized learning as it allows customization so that learners can be addressed on the basis of their requirements at an individual level. To reduce high dropout rates, disengaged learners and curriculum ineffectiveness, such a model can enable learners to guide their own learning with the instructor working more as a mentor/mediator instead of a teacher. The learners can learn at their own pace and can take decisions on their own about their preferences for learning, such as subjects to learn, instructors to learn from, curriculum, patterns and standards to follow being a few of them. It simply means that the educational approach is tailored to her or his individual abilities and needs, which in turn may increase the learner's motivation and likely reduce the dropout rates. In a classroom, there can be two hypothetical scenarios: some learners might experience difficulties/challenges in learning that might require extra tutoring of doubt clearing sessions, while others may be rapidly advancing with their course content and might benefit from additional study materials/assignments that would challenge them on an intellectual level. It would also help the instructors to understand each learner individually and help teach them more effectively.

AI-based learning systems would provide the instructors with information such as the learning style of their learners, their abilities, real-time progress, and information on how to customize their teaching methods for the individual needs of the learners. Machine learning algorithms as explained in the previous section can predict outcomes, which allows providing specific content on the basis of past performance and individual goals of the learner (Deo et al., 2020). For instance, an online learner reflecting a particular skill gap can receive target recommendations that will help in filling the said gap for only that learner. It could alternatively result in the system understanding that the learner may skip a few modules to take a non-linear but comprehensive journey compared to someone who is lacking the basic knowledge of the topic that is being learned. It thereby boosts engagement and results. Also, it allocates the resources to the tasks of value rather than the "one size fits all" methodology. With AI, the LMS may schedule coursework or deliver resources as per the assessment results or simulations, which in turn can help predict the course maps for each learner individually so that they may readjust, if needed. In order to achieve learning objectives of respective courses, such systems have a better return on investment upon comparison with traditional methods. Hence, an administrative perspective suggest spending on such systems over the traditional ones.

Somasundaram et al. reported an AI-enabled intelligent quality management system for the pathway of personalized learning (Somasundaram et al., 2020). They proposed the transformation of teaching and learning processes to intelligent with the power of AI through improvements in curriculum design, planning, delivery and assessment, as shown in Figure 6.3. It can be done on the basis of management information data as an input for data analysis and prediction in order to serve with individual learning plans as personalized learning path.

6.5   INTELLIGENT TUTORING SYSTEMS

An intelligent tutoring system (ITS) is a computer programme that gives tailored instruction on an immediate basis without direct intervention from a human instructor (Steenbergen-Hu & Cooper, 2014). These are computer learning environments capable of helping learners to master skills and knowledge by employing intelligent algorithms that are adaptable at a finer level and represent complex principles of learning by instance. The work done towards development in ITS over the last two decades is focused on two basic pedagogical problems, the first of which is to arrange sophisticated instructional guidance on an individual basis, which is better

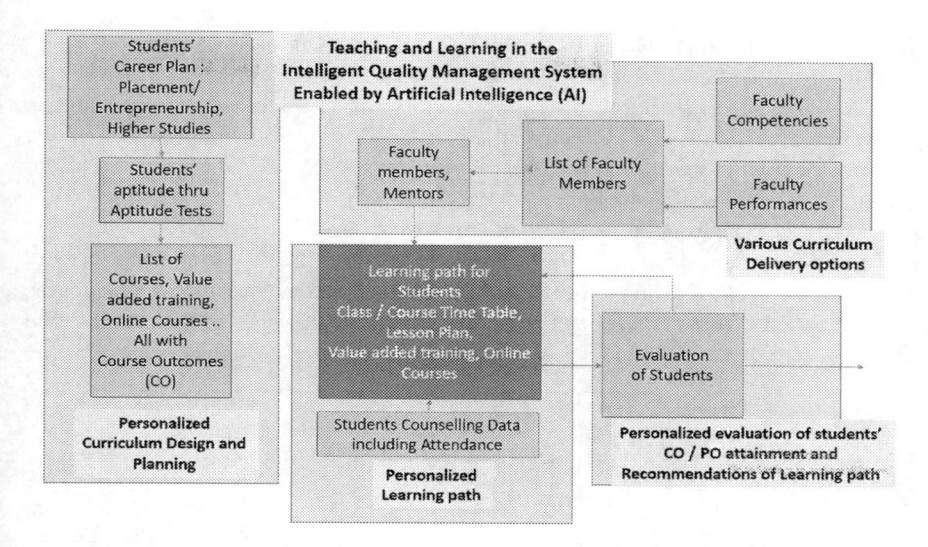

FIGURE 6.3 ML-based Intelligent Quality Management System (Somasundaram et al., 2020).

in terms of results achieved with traditional computer-based instructional systems. The second issue is to proceed with current models on the intellective processes involved with teaching-learning. The research in ITS has leveraged advancements in AI, thereby pushing boundaries of its capabilities with grounded usage scenarios (Kokku et al., 2018). Intelligent Tutoring Systems' (ITSs') cognitive capabilities have been derived from the use of AI techniques. These techniques were put into work with interacting components (Sedlmeier, 2001) such as:

a) The knowledge base, which indicates the centric part of an instructional process;
b) The learner model, indicating the existing state of learner's knowledge;
c) The pedagogical module, indicative of the most suitable instructional approaches that depended upon the learner model assessment; and
d) The user interface, enabling an effective conversation in between the transacting parties—the ITS and a learner.

These components were thoroughly assessed in terms of their granularity, interoperability, reliability and generality, cumulatively indicative of its functionality. ITSs thus can be classified into a) course content curation and delivery ITSs, and b) learner performance monitoring and automated diagnostic feedback offering ITSs. However, issues such as the role of theoretical frameworks in the design, implementation, and validation of an ITS still remain unaddressed (Curilem et al., 2007; Guan et al., 2020; Hwang, 2003).

Web-based ITSs are capable to respond to individual learning needs in pedagogical activities to a greater extent. As learners are involved in education, they should attend educational activities to understand what they are being taught. ITS helps in defining characteristics that in turn help to achieve standards for suggesting activities; for example, responding to some of the learner actions which allow improvement in the speed of learning by adapting to the learners. This web-based adaptive software that responds more to personalized learner needs will comprehend the role of the teachers along with reducing their everyday workload due to their ability to run multiple media and virtual media opportunities. A list of AI and web-based ITSs relevant to the rendering of course content is summarized in Table 6.1.

The development of ITS has been accompanied by three tutoring philosophies, namely: guide, reactive and assistant tutoring. They help in standardizing the degree of intervention of the system. In the guide tutoring philosophy, the system has total

TABLE 6.1

Features of some AI and web-based Intelligent Tutoring Systems.

Sr. No.	Name of ITS	Subject	Observations	Reference
1	ITSMAT	Mathematics	Traditional method for 42 learners resulted in 66% learning performance. ITSMAT improved the performance to 90%.	(Keleş, 2007)
2	PROMATH	Not specified	In addition to four modules of an ITS system, they developed a fifth one, a regulatory module which resulted in elimination of static structures between the instruction and learner modules.	(Dogan & Aktaş, 2011)
3	-	Visual Prolog Language	They created a skeletal system based on information field, user, and user interface models which resulted in learners learning course material more easily and effectively for multiple courses.	(Dag & Erkan, 2004)
4	Negative knowledge-based ITS	English Language	A total of 72 learners evaluated over a 4-week cycle reported significant difference $[t(33) = -7.13$, $p < .05]$ between pre-test and post-test assessments. Upon comparing the experimental and control groups for pretest, no significant differences were observed $[t(65) = -1.15$, $p > .05]$. However, a significant difference $[t(65) = -2.25$, $p < .05]$ was observed in the post-test scores of experimental vs control group learners.	(Demir, 2020)
5	IBITS	Physics-I	In view of academic success and knowledge gained, it resulted in aiding the teacher candidates for learning and comprehending the information about the unit of work, energy and its conservation, which was to be taught to the learners.	(Erdemir & İngeç, 2015)
6	COMET	Medicine	Developed for problem-based learning, this multi-mode interface combined text and graphics in order to have efficient transfer of information between groups and learners. It could be used as a guide for developing new scenarios and human lesson strategies were applied in the cooperative ITS.	(Suebnukarn & Haddawy, 2004)

7	CPP-Tutor	Not Specified	Provided an interactive learning environment with the help of a cover and constraint-based learner model and identification of errors made by the learners with the help of pattern identification for giving feedback. It proved to be a more quick and effective methodology than conventional ones.	(Abu-Naser, 2008)
8	OSCAR-CITS	Not Specified	It aims to mimic a conventional tutor by continually detecting and adapting to an individual's learning styles whilst directing the conversational tutorial. The results exhibited that the learners experiencing a conversational personalized tutorial performed much better than the control group with unmatched tutorials.	(Latham et al., 2014)
9	-	Not Specified	It aims to infer the learner's level of knowledge. A test was employed through which the learner was put and the questions were selected by the system dynamically on the basis of responses given to the previous ones. The relations and probabilistic inference of Bayesian Network could infer many questions and concepts through the evaluation module. It was effective in reinforcing weak topics to cover the learner's needs and provided accurate diagnostic of learner knowledge possession.	(Ramírez-Noriega et al., 2017)
10	Fuzzy based OSCAR CITS	Not Specified	They proposed a new method that employs fuzzy decision trees for constructing fuzzy predictive models combining these variables for all dimensions of the Felder Silverman Learning Styles model. Across four learning style dimensions, fuzzy models have improved the predictive accuracy of OSCAR CITS and have exhibited some interesting relationships between behaviour variables.	(Crockett et al., 2017)
11	JavaTutor	Introductory Computer Science	Helps the learners with both cognitive and affective feedback which provides an interactive and scalable learner support system. The novel framework with integrated behaviours were capable to leverage ML for acquiring tutorial strategies from data collected within novice learners and experienced human tutors.	(Wiggins et al., 2015)
12	-	English Grammar	Designed a ITS for teaching English language grammar in order for them to learn easily and smoothly. They employed the use of an ITS builder tool for learning grammar tenses using Delphi Embarcadero XE8, 2015 which comprises of two sub systems as teacher and learner to add the course content, assessment questions, and answers; and to learn the course material along with answering to the exercises, respectively.	(Alhabbash et al., 2016)

control over the process and interactions of the learner, and the systems guide him or her towards gaining new knowledge. The reactive tutoring philosophy allows the learner to have control over the interactions and the system reacts to the learner's actions. The third tutoring philosophy or the assistant method uses a Learner Model and associated strategies to automatically adjust the system interventions on the basis of the learner. In modern-day ITSs, researchers look to integrate all three philosophies so that the system is well equipped with various strategies for learning to be effective and efficient at the same time. In all implementations, there is a learner model, and the tutor module preserves the knowledge that allows for decision-making regarding configuring the output on the basis of input of learner and models (Curilem et al., 2007).

6.5.1 ITS Architecture

The architecture of an ITS system comprises a knowledge base, learner model, pedagogical module and a user interface model (Almasri et al., 2019). The source of knowledge to the learners, also known as a knowledge base or the domain model, relates to the course in which the learner is expected to learn, answer questions and solve problems related to the same (Ma et al., 2014). The learner model represents the processes that function on the knowledge such as problem-solving, retrieval of information, learning from mistakes, level of individual learning, and patterns of learning (Brown, 2009). The tutoring or pedagogical module comprises knowledge about how to educate learners and rules of particular course instructions (Polson & Richardson, 2013). Lastly, the user interface model is the learner interaction with ITS, in order to ascertain the dialogues and learner communication with ITS (Qwaider & Abu-Naser, 2018). The ITS architecture can be represented as given in Figure 6.4.

6.5.2 Applications of Intelligent Tutoring Systems

To follow are listed some of the globe's leading ITS with well-established clientele. We have tried to summarize what they are offering in the form of an ITS product with the information published on their webpages.

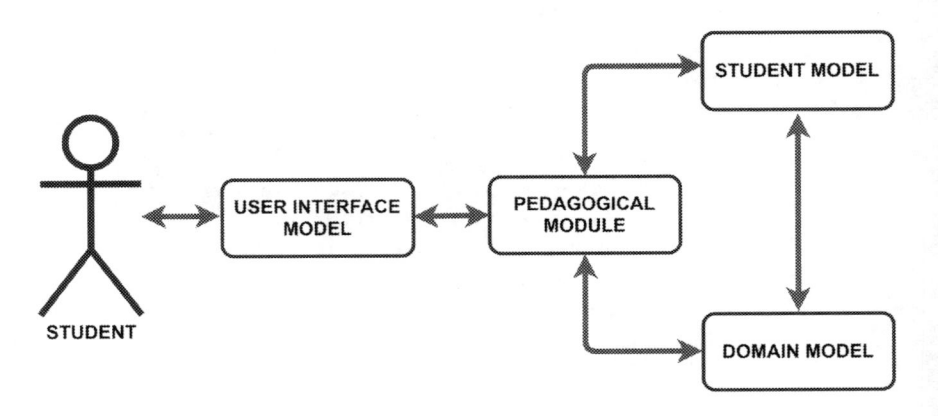

FIGURE 6.4 Architecture for an Intelligent Tutoring System.

6.5.2.1 DreamBox (www.dreambox.com/)

It is a digital programme designed for learning mathematics curriculum with rigorous and interactive lessons that can adapt to individual learner needs in order to provide an ultimate personalized learning experience. This enables in both assisting and challenging the learner at the same time, irrespective of where the learner has started, thus assisting the teachers as a teaching assistant. Some of the unique features such as lesson recommendations give educators valuable information about learning thereby enabling the educator with a quick understanding of the state and position of the learner so they may explore if there is a possibility to group certain learners with respect to their positions in order to bring exploratory group learning into the picture.

6.5.2.2 ALEKS (www.aleks.com/)

It is a research-based, online programme based on ANN that offers learning in mathematics and science-related courses with a history of 20 years in research and analytics. It is efficient in pinpointing exactly the knowledge state of the learner in order to suggest personalized learning pathways, so that every learner may develop mastery over a course irrespective of their starting point. They start with an individualized assessment with questions chosen using learning space theory based on the learner's response to the previous questions asked. Upon completion, the knowledge state of the learner is visible in the form of a multi-coloured pie chart wherein the learner understands what topics they have mastered, not yet mastered, and are ready to learn, along with topics available to explore for learning. The system continually records the knowledge state of the learner through the assessment carried out during the learning process. Also, periodic assessments help to keep track of the knowledge retention of the topics that the learner has passed through.

6.5.2.3 Imagine Learning Language Advantage (www.imaginelearning.com/math)

These are supplemental language learning programs that provide adaptive and age-appropriate learning environments for mathematics learners. Language is considered to be a key element in mathematics. While we mostly compute numbers in mathematics, language is utilized to discuss it along with the theory, approaches, contexts, and more. Their systems provide K–12 instructors with an accurate measurement of growth, mastery and prediction of state test performance.

6.5.2.4 Knewton (www.knewton.com/)

Knewton develops personalizing technology for educational purposes. Alta is their newest product in the form of a complete courseware solution that has excellent adaptive learning technology with quality course content in order to offer a personalized learning experience that can improve learner outcomes in an affordable and accessible manner for mathematical courses such as statistics, chemistry, economics and so on. It is able to quickly identify and boost knowledge gaps dynamically while completing assignments. Moreover, it's a mobile-based technology that is available on the go. The Knewton factor in Alta, enabling access to a database of over 15 million learners worldwide, helps in easy identification of knowledge gaps and dynamically giving instructions to fill them, while one is completing coursework. With Alta, the assignments are so personalized and based on one's knowledge levels that it intrigues a learner to proceed and not leave it incomplete.

6.6 CASE STUDIES

Mousavinasab et al. summarized the characteristics, applications and evaluation methods for ITSs (Mousavinasab et al., 2021). Of the carefully selected papers for review from 2007 to 2017, they reported the highest use of ITS in computer science and engineering followed by mathematics and health/medical with 38% in computer science and 15% each in other two respectively, as shown in Figure 6.5. Around 75% of the observed literature involved subjects who were university learners.

They also reported the use of AI techniques such as ANN, Data mining, Fuzzy based, NLP based, case-based, condition action rule-based reasoning, Bayesian-based techniques, and intelligent agents being used together or on an individual basis, with condition action rule-based reasoning having the highest percentage of frequencies at 34%, followed by data mining at 23% and Bayesian-based techniques at 21% closing the three highest frequencies reportedly employed. In computer science and mathematics courses, condition action rule-based reasoning was effective for designing ITS related to them. The user interface was majorly web-based with 55% client computers, and 15% were mobile-based. Furthermore, a host of reasons for the use of applying AI techniques were established of which presenting learning material or content and adaptive learning path navigation, adaptive feedback or recommendation generation had a frequency of 41.5%, 28%, and 52.83%, respectively. They identified the characteristics helping to deliver effective adaptive learning as knowledge level (62%), behaviour in learning path (41.5%) and learning performance (52.83%) with other engaging ones such as learning preference, learning style, cognitive factors,

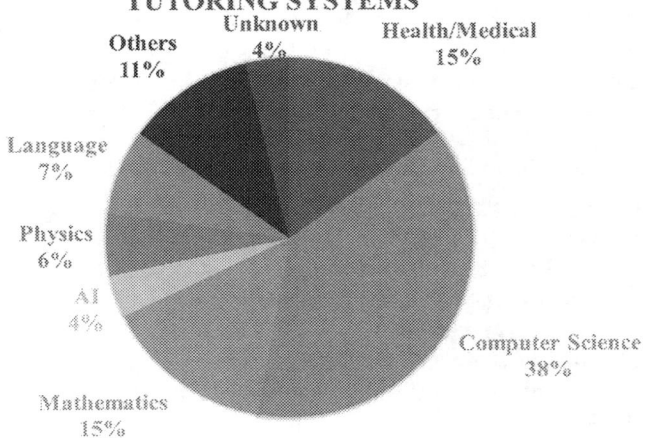

FIGURE 6.5   Frequency of Educational Areas in ITSs (Mousavinasab et al., 2021).

emotional factors, cultural factors, intelligence level as per decreasing observation (frequency) of occurrence in the literature review. To sum it up, they categorized ITSs based on AI techniques and put them into six groups as shown in Table 6.2 with sub-items in each group defining the sub purpose of classification and use.

ITS should respond to the learner's individual needs by automatically generating compatible courses, their content and providing responses according to the history of the learner's profile. The grueling part in doing such development is the collection and codification of the knowledge domain and establishing its connections with each other. With respect to solving this issue, Zrigui and Moussa developed an ITS built on an interface capable of offering the tutor a possibility to add didactical and associated pedagogical knowledge, respectively (Zrigui & Moussa, 2017). They described the system architecture exhibiting the organization of knowledge basis and employed a method for automated insertion of the course content in the knowledge domain. The effort was focused on the reduction of the design and implementation complexity which saves valuable cost and time. They aimed to transform an input course form into an XML code keeping the representation of knowledge for the ITS models. Also, it adds new content to the knowledge base.

Schiffaino et al. developed an intelligent teaching agent called eTeacher, built to assist e-learners in a personalized manner and tested for Systems Engineering course (Schiaffino et al., 2008). The learner profile was built with the help of observations

TABLE 6.2

Categorization of Overall Need of Applying AI Techniques in ITSs (Mousavinasab et al., 2021).

Sr. No.	Purpose Item	Purpose Sub items
1	Adaptive Guidance	• Adaptive feedback generation • Adaptive hint generation • Adaptive recommendation generation
2	Adaptive Instruction	• Rendering adaptive course Content • Adaptive learning path navigation • Presenting adaptive test and exercises
3	Learner's Evaluation	• Evaluation of Knowledge • Evaluation of Performance • Evaluation of Skills
4	Define and update the learner's model based on	• Style of Learning • Level of Knowledge
5	Classify/Cluster learners on the basis of	• Affect • Intelligence • Style of Learning • Needs of Learning • Learning Characteristics
6	Others	• Communication • Calculation of difficulty level of exercises • Learning material classification

in a learner's behaviour while they take online classes. The profile includes learning style, performance assessment for exercises done, topics undertaken for study, results of exam conducted, and so on. Learning style is automatically perceived through actions using Bayesian networks which, along with other information in the learner profile, assists the eTeacher to engage and help the learner on a proactive basis by suggesting them personalized courses that will assist them during the process of learning. The interdependencies between the learning styles and behaviours are encoded in the Bayesian model as shown in Figure 6.6 via the arcs that travel from the nodes indicative of learner behaviour to the nodes indicative of learning style dimensions. It is also indicative of a conditional probability table for the understanding node.

Cardoso et al. proposed a multi-agent intelligent tutoring system building tool that facilitates the task of a teacher in content development of a tutorial course system along with providing flexibility and adaptiveness in its presentation (Cardoso et al., 2004). This facilitation with the help of adopted formalisms are ground logic terms for the learner model, databases for the domain model, and object Petri nets for the methodological model. The learner-agent system interaction is commandeered by an object Petri net, which gets auto-translated in an expert system that is rule-oriented. The object Petri net tokens comprise of data objects that possess pointers to the learner model and the domain knowledge which is stored into text, example, and exercise form database. The object Petri net transitions are commandeered by logical conditions pertaining to the learner model and its firing produces actions that update it. Zhuge and Li proposed an authoring model for the development of ITSs in a very effective and pleasant way (Zhuge & Li, 2004). The model, KGTutor, a knowledge grid-based ITS, has shown promise to support the decentralized, learner-centric, and highly interactive approach. The crux of KGTutor was to organize the course into a concept space instead of page space. The system utilized learner characteristics such as background knowledge, learning styles and so on for selection, organization, and delivery of course learning materials in a personalized manner to individual learners. The system was also able to provide assessments in an objective manner along with offering personalized learner suggestions on the basis of their learning performance.

6.7   CHALLENGES AND DRAWBACKS

The last decade has seen the emergence of a number of robust tools in the maturation of statistical and computational methods for predictive modelling for educational teaching and learning data. However, issues are faced during building, validating and applying predictive models. In order for the predictive modelling techniques to have the impact of their potential, some of the areas may require investments in order to do so (Brooks et al., 2010). One such area is the support to be provided to educational researchers, cognitive and social psychologists/behaviourists, psychometricians and policy experts in predictive modelling techniques. This group of people are good at exploratory modelling and thus provision of support for applications of predictive modelling techniques, either by innovation through user-friendly tools or educational resources development on predictive modelling. The second area is the creation of community-led educational data science challenge initiatives. The same theme of work can be addressed with the use of different datasets, their approach/

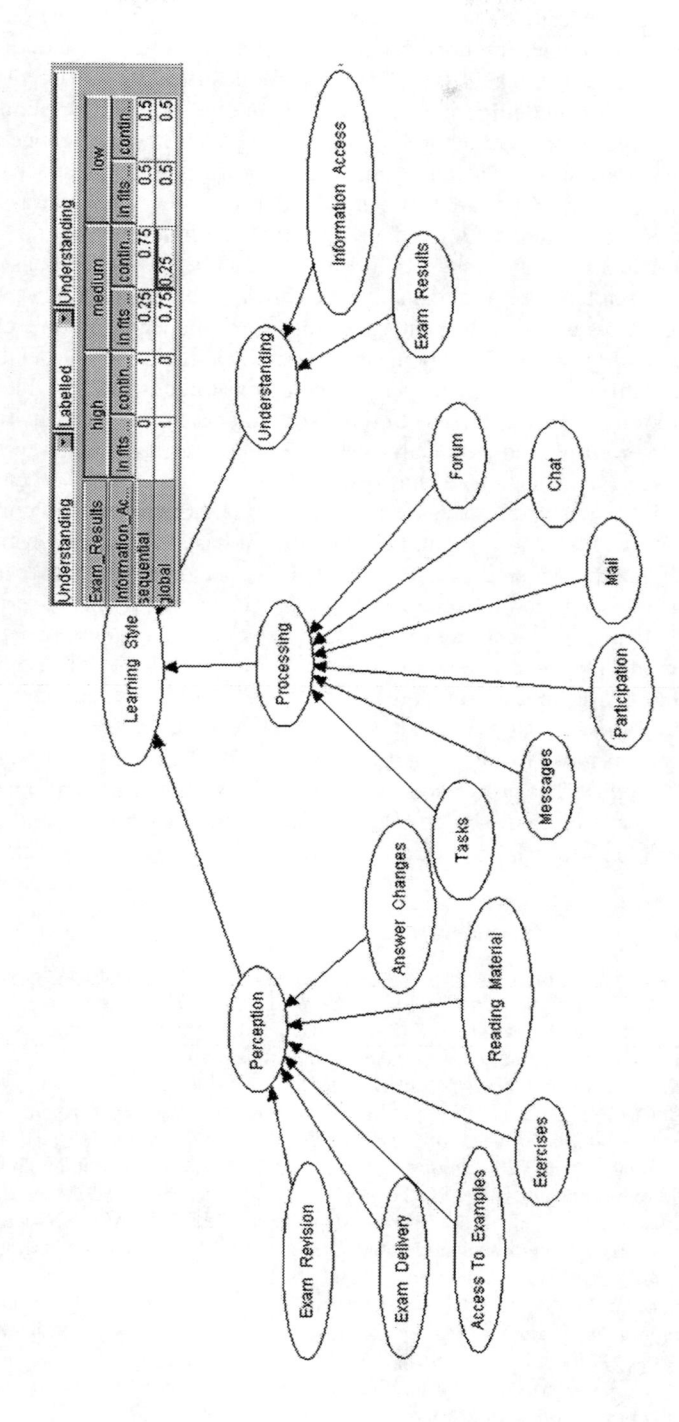

FIGURE 6.6 Bayesian Network for Detection of Learning Styles (Schiaffino et al., 2008).

implementation and subsequent outcomes variables. These results become difficult to compare and arrive at a general conclusion, which from the societal point of view can be detrimental, as it would be difficult to compare the efficacy of different techniques and suitability of modelling methods. Thirdly, it can be support extended to engage in second order predictive modelling. In learning analytics, the second order predictive models are the ones including historical knowledge with respect to the effects of and intervention of the same model. A model accessing drop out ratio with the help of learner interactions with the course content is a first order model, while the one which has an intervention, such as an automated mailer, would be considered a second order one. In the case of personalized learning paths, understanding intervention effectiveness would be important, especially when there are multiple ones.

Education technology is not just about technology; it is inclusive of social, cultural, economic, ethical, pedagogical and psychological dimensions of education. AI-based education is interdisciplinary. In the last decade or so, the introduction of Big Data, learner profiling and personalized learning design has enabled learning analytics to become a centric area. AI has been successfully employed in the development of ITSs. It has enabled learners to recognize their shortfalls, provide instant feedback and learn at different levels at the same time. It has also reduced instructor work in planning and has catered to individualized and personalized learner needs. They can readily be integrated into classes to provide adequate support to teachers. However, the challenge for the system is to be adaptive as per the system requirements. Also, simplifying the architecture by reducing its complexity of design and implementation of ITS is an important area to look in the future. Other factors such as limited training time, constraints with educational curriculum, and pedagogical pairing limitations may also result in challenges (Cakir, 2019). The use of ML methods such as ANN and DL has proven to be very useful in predictive modelling, along with algorithms like action rule-based, Bayesian network, data mining and so on leading the way in ITSs for adaptive learning systems.

REFERENCES

Abu-Naser, S. S. (2008). Developing an intelligent tutoring system for learners learning to programme in C++. *Information Technology Journal*, 7(7), 1055–1060.

Alhabbash, M. I., Mahdi, A. O., & Naser, S. S. A. (2016). An intelligent tutoring system for teaching grammar English tenses. *European Academic Research*, 4(9).

Almasri, A., Ahmed, A., Almasri, N., Abu Sultan, Y. S., Mahmoud, A. Y., Zaqout, I. S., Akkila, A. N., & Abu-Naser, S. S. (2019). Intelligent tutoring systems survey for the period 2000–2018. *International Journal of Academic Engineering Research*, 3(5), 21–37.

Alshanqiti, A., & Namoun, A. (2020). Predicting learner performance and its influential factors using hybrid regression and multi-label classification. *IEEE Access*, 8, 203827–203844.

Asiah, M., Zulkarnaen, K. N., Safaai, D., Hafzan, M. Y. N. N., Saberi, M. M., & Syuhaida, S. S. (2019). A review on predictive modeling technique for learner academic performance monitoring. *MATEC Web of Conferences*, 255, 03004.

Baker, T., Smith, L., & Anissa, N. (2019). *Educ-AI-tion rebooted? Exploring the future of Artificial Intelligence in schools and colleges* (pp. 1–56) [Project Report]. Nesta. https://media.nesta.org.uk/documents/Future_of_AI_and_education_v5_WEB.pdf.

Benke, K., & Benke, G. (2018). Artificial Intelligence and big data in public health. *International Journal of Environmental Research and Public Health*, 15(12), 2796.

Boobier, T. (2020). *AI and the future of banking.* Chichester: John Wiley & Sons.

Brooks, C., & Thompson, C. (2017). Predictive modelling in teaching and learning. *Handbook of Learning Analytics,* 61–68.

Brooks, C., Thompson, C., Ri, H. D., Hgxfdwlrqdo, D., & Prghoolqj, S. (2010). Chapter 5: Predictive modelling in teaching and learning. *Handbook of Learning Analytics,* 61–68.

Brooks, C., Thompson, C., & Teasley, S. (2015). *A time series interaction analysis method for building predictive models of learners using log data.* Proceedings of the 5th International Conference on Learning Analytics and Knowledge, pp. 126–135.

Brown, Q. (2009). *Mobile intelligent tutoring system: Moving intelligent tutoring systems off the desktop.* Philadelphia, PA: Drexel University.

Cakir, R. (2019). Effect of web-based intelligence tutoring system on learners' achievement and motivation. *Malaysian Online Journal of Educational Technology,* 7(4), 45–59.

Cao, L. (2020). *AI in finance: A review.* Available at SSRN 3647625.

Cardoso, J., Bittencourt, G., Frigo, L. B., Pozzebon, E., & Postal, A. (2004). *MathTutor: A multi-agent intelligent tutoring system.* IFIP International Conference on Artificial Intelligence Applications and Innovations, pp. 231–242.

Chien, Y.-C., Liu, M.-C., & Wu, T.-T. (2020). Discussion-record-based prediction model for creativity education using clustering methods. *Thinking Skills and Creativity,* 36, 100650.

Ciolacu, M., Tehrani, A. F., Binder, L., & Svasta, P. M. (2018). *Education 4.0-Artificial Intelligence assisted higher education: Early recognition system with machine learning to support learners' success.* 2018 IEEE 24th International Symposium for Design and Technology in Electronic Packaging (SIITME), pp. 23–30.

Crockett, K., Latham, A., & Whitton, N. (2017). On predicting learning styles in conversational intelligent tutoring systems using fuzzy decision trees. *International Journal of Human-Computer Studies,* 97, 98–115.

Curilem, S. G., Barbosa, A. R., & de Azevedo, F. M. (2007). Intelligent tutoring systems: Formalization as automata and interface design using neural networks. *Computers & Education,* 49(3), 545–561. https://doi.org/10.1016/j.compedu.2005.10.005.

Dag, F., & Erkan, K. (2004). Prolog based intelligent tutoring systems (ITS). *Pamukkale University Journal of Engineering Sciences,* 10(4), 47–55.

Demir, Ü. (2020). The effect of using negative knowledge based intelligent tutoring system evaluator software to the academic success in English language education. *Pedagogies: An International Journal,* 15(4), 245–261. https://doi.org/10.1080/1554480X.2019.1706522.

Deo, R. C., Yaseen, Z. M., Al-Ansari, N., Nguyen-Huy, T., Langlands, T. A. M., & Galligan, L. (2020). Modern Artificial Intelligence model development for undergraduate learner performance prediction: An investigation on engineering mathematics courses. *IEEE Access,* 8, 136697–136724. https://doi.org/10.1109/ACCESS.2020.3010938.

Deshmukh, S. V. (2018). Artificial Intelligence in dentistry. *Journal of the International Clinical Dental Research Organization,* 10(2), 47.

Dogan, N., & Aktaş, B. (2011). Promath: Web Tabanlı Zeki Öğretim Sistemleri İçin Düzenleyici Modül Uygulaması. *Bilişim Teknolojileri Dergisi,* 4(2).

Erdemir, M., & İngeç, Ş. K. (2015). The influence of web-based intelligent tutoring systems on academic achievement and permanence of acquired knowledge in physics education. *US-China Education Review A,* 5(1), 15–25.

Garg, S. (2021). Intelligent pattern classification of factors affecting concurrent engineering techniques. *2021 7th International Conference on Advanced Computing and Communication Systems (ICACCS),* 1, 1864–1871.

Georgescu, A.-A. (2018). Chatbots for education—trends, benefits and challenges. *Conference Proceedings of ELearning and Software for Education "(ELSE),* 2(14), 195–200.

Gladshiya, V. B., & Sharmila, D. K. (2019). A review study on predictive analytical tools and techniques in education. *International Journal of Engineering Research & Technology, 8*(11). Available at: www.ijert.org/research/a-review-study-on-predictive-analytical-tools-and-techniques-in-education-IJERTV8IS110411.pdf, www.ijert.org/a-review-study-on-predictive-analytical-tools-and-techniques-in-education.

Gray, G., McGuinness, C., & Owende, P. (2014). An application of classification models to predict learner progression in tertiary education. *2014 IEEE International Advance Computing Conference (IACC)*, 549–554.

Guan, C., Mou, J., & Jiang, Z. (2020). Artificial Intelligence innovation in education: A twenty-year data-driven historical analysis. *International Journal of Innovation Studies, 4*(4), 134–147.

Hasan, R., Palaniappan, S., Mahmood, S., Sarker, K. U., & Abbas, A. (2020). Modelling and predicting learner's academic performance using classification data mining techniques. *International Journal of Business Information Systems, 34*(3), 403–422. https://doi.org/10.1504/IJBIS.2020.108649.

Huang, M.-H., & Rust, R. T. (2018). Artificial Intelligence in service. *Journal of Service Research, 21*(2), 155–172.

Huang, S., & Fang, N. (2010). Regression models for predicting learner academic performance in an engineering dynamics course. *2010 Annual Conference & Exposition*, 15–1026.

Hwang, G.-J. (2003). A conceptual map model for developing intelligent tutoring systems. *Computers & Education, 40*(3), 217–235. https://doi.org/10.1016/S0360-1315(02)00121-5.

Jarrahi, M. H. (2018). Artificial Intelligence and the future of work: Human-AI symbiosis in organizational decision making. *Business Horizons, 61*(4), 577–586.

Keleş, A. (2007). *Öğrenme-öğretme sürecinde yapay zekâ ve web tabanlı zeki öğretim sistemi tasarımı ve "matematik öğretiminde bir uygulama*. Unpublished Dissertation, Atatürk Üniversitesi, Erzurum.

Khazanchi, R., & Khazanchi, P. (2021). Artificial Intelligence in education: A closer look into intelligent tutoring systems. In *Handbook of research on critical issues in special education for school rehabilitation practices* (pp. 256–277). Hershey, PA: IGI Global.

Knowledge Works. (2013, July 1). A glimpse into the future of learning: An infographic. *A Glimpse into the Future of Learning*. Available at: https://knowledgeworks.org/resources/future-learning-infographic/.

Kokku, R., Sundararajan, S., Dey, P., Sindhgatta, R., Nitta, S., & Sengupta, B. (2018). *Augmenting classrooms with AI for personalized education*. 2018 IEEE International Conference on Acoustics, Speech and Signal Processing (ICASSP), pp. 6976–6980. https://doi.org/10.1109/ICASSP.2018.8461812.

Latham, A., Crockett, K., & McLean, D. (2014). An adaptation algorithm for an intelligent natural language tutoring system. *Computers & Education, 71*, 97–110. https://doi.org/10.1016/j.compedu.2013.09.014.

Leyva-Vázquez, M., Smarandache, F., & Ricardo, J. E. (2018). Artificial Intelligence: Challenges, perspectives and neutrosophy role (Master conference). *Dilemas Contemporáneos: Educación, Política y Valore, 6*(Special).

Liu, S., Chen, Y., Huang, H., Xiao, L., & Hei, X. (2018). *Towards smart educational recommendations with reinforcement learning in classroom*. 2018 IEEE International Conference on Teaching, Assessment, and Learning for Engineering (TALE), pp. 1079–1084. https://doi.org/10.1109/TALE.2018.8615217.

Ma, W., Adesope, O. O., Nesbit, J. C., & Liu, Q. (2014). Intelligent tutoring systems and learning outcomes: A meta-analysis. *Journal of Educational Psychology, 106*(4), 901.

Mou, X. (2019). *Artificial Intelligence: Investment trends and selected industry uses* (Brief No. 143357, pp. 1–8). IFC, A member of the world bank group. https://documents.worldbank.org/en/publication/documents-reports/documentdetail/617511573040599056/artificial-intelligence-investment-trends-and-selected-industry-uses.

Mousavinasab, E., Zarifsanaiey, N., R. Niakan Kalhori, S., Rakhshan, M., Keikha, L., & Ghazi Saeedi, M. (2021). Intelligent tutoring systems: A systematic review of characteristics, applications, and evaluation methods. *Interactive Learning Environments*, 29(1), 142–163.

Polson, M. C., & Richardson, J. J. (2013). *Foundations of intelligent tutoring systems*. New York: Psychology Press.

Qwaider, S. R., & Abu-Naser, S. S. (2018). Excel intelligent tutoring system. *International Journal of Academic Information Systems Research*, 2(2), 8–18.

Ramírez-Noriega, A., Juárez-Ramírez, R., & Martínez-Ramírez, Y. (2017). Evaluation module based on Bayesian networks to intelligent tutoring systems. *International Journal of Information Management*, 37(1 Part A), 1488–1498. https://doi.org/10.1016/j.ijinfomgt.2016.05.007.

Rastrollo-Guerrero, J. L., Gómez-Pulido, J. A., & Durán-Domínguez, A. (2020). Analyzing and predicting learners' performance by means of machine learning: A review. *Applied Sciences*, 10(3), 1042. https://doi.org/10.3390/app10031042.

Sandu, N., & Gide, E. (2019). *Adoption of AI-Chatbots to enhance learner learning experience in higher education in India*. 2019 18th International Conference on Information Technology Based Higher Education and Training (ITHET), 1–5.

Schiaffino, S., Garcia, P., & Amandi, A. (2008). eTeacher: Providing personalized assistance to e-learning learners. *Computers & Education*, 51(4), 1744–1754. https://doi.org/10.1016/j.compedu.2008.05.008.

Sedlmeier, P. (2001). Intelligent tutoring systems. In N. J. Smelser & P. B. Baltes (Eds.), *International encyclopedia of the social & behavioral sciences* (pp. 7674–7678). Pergamon. https://doi.org/10.1016/B0-08-043076-7/01618-1.

Shawky, D., & Badawi, A. (2018). A reinforcement learning-based adaptive learning system. In A. E. Hassanien, M. F. Tolba, M. Elhoseny, & M. Mostafa (Eds.), *The international conference on advanced machine learning technologies and applications (AMLTA2018)* (pp. 221–231). Springer International Publishing. https://doi.org/10.1007/978-3-319-74690-6_22.

Somasundaram, M., Junaid, K. A. M., & Mangadu, S. (2020). Artificial Intelligence (AI) enabled intelligent quality management system (IQMS) for personalized learning path. *Procedia Computer Science*, 172, 438–442. https://doi.org/10.1016/j.procs.2020.05.096.

Sood, S., & Saini, M. (2021). Hybridization of cluster-based LDA and ANN for learner performance prediction and comments evaluation. *Education and Information Technologies*, 26(3), 2863–2878.

Steenbergen-Hu, S., & Cooper, H. (2014). A meta-analysis of the effectiveness of intelligent tutoring systems on college learners' academic learning. *Journal of Educational Psychology*, 106(2), 331–347. https://doi.org/10.1037/a0034752.

Suebnukarn, S., & Haddawy, P. (2004). A collaborative intelligent tutoring system for medical problem-based learning. *Proceedings of the 9th International Conference on Intelligent User Interfaces*, 14–21.

Tsai, S.-C., Chen, C.-H., Shiao, Y.-T., Ciou, J.-S., & Wu, T.-N. (2020). Precision education with statistical learning and deep learning: A case study in Taiwan. *International Journal of Educational Technology in Higher Education*, 17(1), 12. https://doi.org/10.1186/s41239-020-00186-2.

Wang, Y. (2021). When Artificial Intelligence meets educational leaders' data-informed decision-making: A cautionary tale. *Studies in Educational Evaluation*, 69, 100872.

Wiggins, J. B., Boyer, K. E., Baikadi, A., Ezen-Can, A., Grafsgaard, J. F., Ha, E. Y., Lester, J. C., Mitchell, C. M., & Wiebe, E. N. (2015). JavaTutor: An intelligent tutoring system that adapts to cognitive and affective states during computer programming. *Proceedings of the 46th ACM Technical Symposium on Computer Science Education*, 599. https://doi.org/10.1145/2676723.2691877.

Zhuge, H., & Li, Y. (2004). KGTutor: A knowledge grid based intelligent tutoring system. In J. X. Yu, X. Lin, H. Lu, & Y. Zhang (Eds.), *Advanced web technologies and applications* (pp. 473–478). Springer. https://doi.org/10.1007/978-3-540-24655-8_51.

Zrigui, S., & Moussa, A. A. (2017). Automatic insertion of subject content in domain knowledge of an intelligent tutoring system. *2017 Intelligent Systems and Computer Vision (ISCV)*, 1–5.

7 Impact of AI on Teaching Pedagogy and its Integration for Enhancing Teaching-Learning

Bhagwan Toksha, Trishul Kulkarni and Prashant Gupta

CONTENTS

7.1 INTRODUCTION

The itinerant progress of humanity has taken new pathways in the 21st century. The education system is a major catalyst in societal changes. It is coping with the evolving needs, changing definitions and requirements from the professional perspective as well as overall living styles. The focus is shifting from compliance and conformity towards skill sets and creative mindsets. The ethics, values and basic framework of teaching-learning still remain intact, while methods to discover and nurture learners' talent of towards broader interests are occupying a larger space. Today's society needs innovation and creativity in many areas going beyond the framework of standardized curriculum through conventional learning and a "one size fits all" pace type of assessment and evaluation.

Fortunately, progress is being made to develop faster rates of information exchange and more availability of digitized, interactive content, rather than clinging to obsolete methods. The education system has responded openly to technological developments such as Information and communication technologies (ICT) and more

DOI: 10.1201/9781003184157-7

recently to Artificial Intelligence (AI). There is a fine line differentiating education technology and Artificial Intelligence in education. In this view, the facilities AI possibly extend towards education along with the possible benefits and challenges are summarized in Figure 7.1.

The effect of rapidly developing Artificial Intelligence (AI) technologies will certainly pave the way for newer, enriching, profound, yet sometimes overhyped,

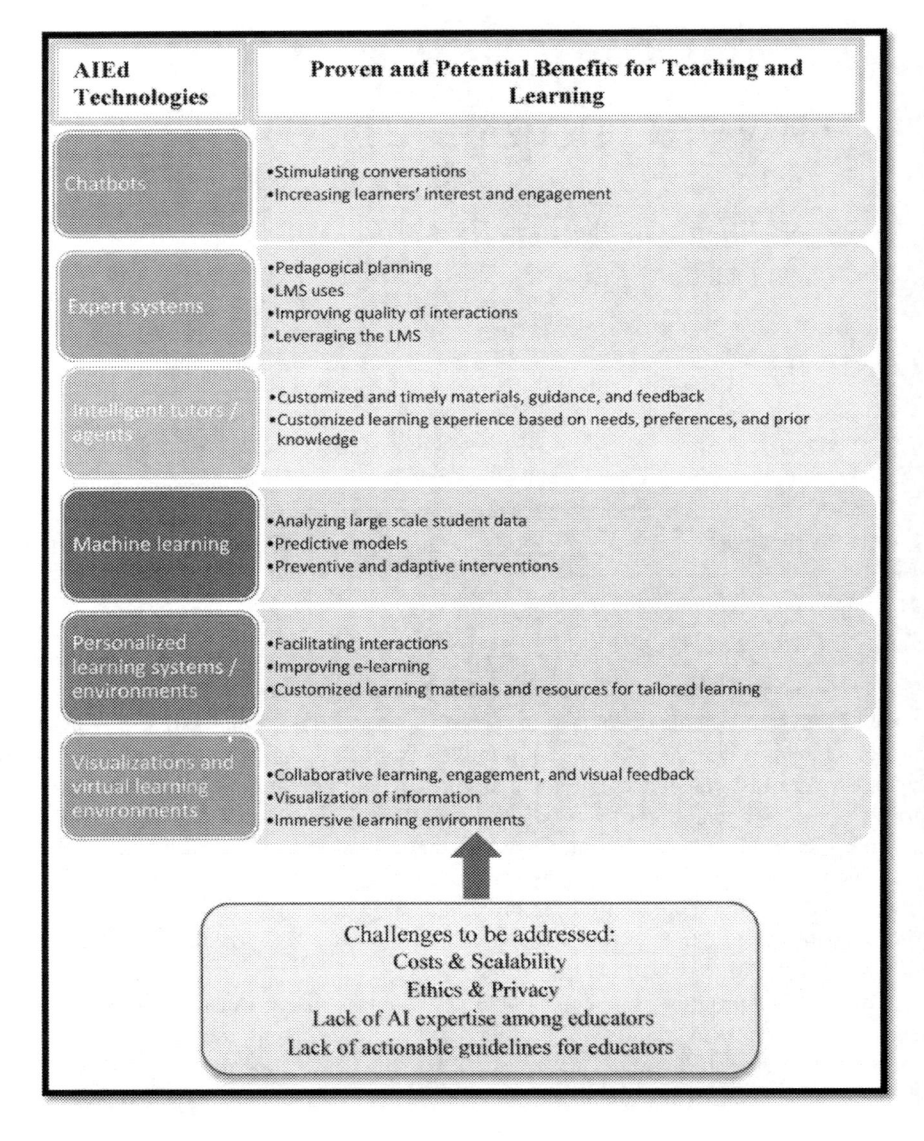

FIGURE 7.1 The Facilities Extended from AI Towards Education, Their Possible Benefits and Challenges Adopted from Zhang and Aslan, 2021 (Zhang & Aslan, 2021).

possibilities regarding education. These technologies are generating a growing interest about their application in educational contexts. The keyword "teaching/training" spans traditional classrooms as well as skill transfer at workplaces and supports conventional educational setup as well as lifelong learning. The very nature of AI and learning science is interdisciplinary (Broggy et al., 2017; Zhuang et al., 2020). AI is evolving in the terrains of psychological and cognitive modelling, mind philosophy, and computer science terrain such as data acquisition, training, testing and deployment.

The learning sciences encompass fields such as education, psychology, neuroscience, linguistics, sociology and anthropology. These fields, when joined, would promote the development of smart learning ecosystems that are flexible, inclusive, personalized, engaging and effective. The 'pedagogical approach' to teaching is a phrase, which has dimensions of the method, and practice, of teaching. It includes teaching theories, teaching styles and feedback/assessment mechanisms. Generally, the concept of pedagogy regards the way instructors deliver content to students. The lesson plan design depends on how the instructor chooses to deliver the content. The choices may be based on the instructors' personal teaching preferences, experience and the setup context/environment. Technological developments have changed the reasons why a learner would be interested in learning. The learner may be not seeking to gain knowledge via learning as the information is already at his/her fingertips in the best possible and presentable form. The pace at which the learner understands the content would be greatly enhanced by the availability of interactive/automated virtual platforms for performing experiments, enabling learners to apply the content and higher order skills.

The work of testing compatibility of AI technologies and pedagogies has a good amount of history (Woolf, 1988). Considerable investments are made in the creation of AI-based education solutions. There are new establishments coming such as Alta for higher education, Knewton (www.knewton.com/, 2021) and K-12 from Carnegie Learning (www.carnegielearning.com, 2021), Cognilytica (www.cognilytica.com/, 2018) and netex (www.netexlearning.com/en/, 2019), with the products enabling learners to take control of their own learning. AI-based education frameworks are projected to enhance teaching, learning and teacher training (Vlasova et al., 2019; Rita Liao, 2018). The point of focus is that the application of AI in educational contexts is inevitable and expanding at a quick rate, such that in 2020, total investments were at \$16 billion (Ascione, 2017; www.holoniq.com/, 2021). Considering this situation, it was planned to present the various aspects, pedagogical approaches and AI integration, and visit the design of pedagogical solutions based on AI in the form of this chapter. The role that Artificial Intelligence is playing in pedagogy and the teaching-learning processes along with the possible benefits of integrating AI in pedagogy and possible concerns and challenges are also discussed.

7.2 PEDAGOGICAL APPROACHES AND AI INTEGRATION

Educational pedagogies and relevant technologies are distinct from pure technologies in a sense that these must be human-centred. The very reason for this is the involvement of design and deployment of instructional and interactive environments. Thus, AI-enabled intelligent learning pedagogies must go beyond performance and focus on human feelings and learning outcomes. The medical jargon such as precise, site-specific drug delivery came as a motivation in the inclusion of AI in education (Duong et al., 2019; Kuch et al., 2020). Generally, such precise and site-specific needs exceed beyond the "one size fits all" approach, as learning styles differ greatly, due to causes including but not limited to biomedical, sociocultural surroundings and lifestyles (Lin et al., 2021). This creates the need for individualized and precise pedagogical steps: diagnosis, forecasting, treatment and prevention of failure. The learner needs help to face learning challenges, like poor performance and learning disabilities. AI-enabled pedagogies will consider the learners' learning behaviours, learning environments, and implemented strategies and provide effective solutions in all four departments of diagnosis of individual problems, forecasting, treatment and prevention of failure (Yang, 2021).

AI techniques encompass (and are not limited to) backing up and contributing to teaching/learning, the design, development of e-learning systems and pedagogical innovations. AI performing various roles as pedagogical tools in teaching/learning from the literature are delineated in Table 7.1. AI also finds applications in educational data mining, maintaining records and learning analytics. Moreover, the advancements in AI techniques would be helpful for domain-specific applications in education, educational data analysis and learner modelling. The instructional design model ADDIE supported by AI will reduce a considerable labour involved in terms

TABLE 7.1
The Delineations of AI as Pedagogical Tools in Teaching/Learning.

Sr. No	Delineation	Authors
1	AI as an intelligent tutoring system assisting to organize system knowledge and operational information to enhance operator performance and automatically determine progression and remediation during a training session, by referring to past student performance.	Hwang 2003 (Hwang, 2003)
2	AI as an artificially intelligent tutor that provides personalized instructions and explanations throughout the learning track.	Johnson et al. 2009 (Johnson et al., 2009)
3	AI as a pedagogical assistant in STEM courses with a blend of inherent AI features and cognitive science.	Benedict du Boulay 2016 (du Boulay, 2016)
4	AI as a robotic entity assistant to the instructor supporting assessment in online environments.	Popenici and Kerr 2017 (Popenici & Kerr, 2017)
5	AI as a communication agent with virtues such as social ambiance, celerity and wittiness. Instructors modified its role more towards monitoring learner's progress.	Edwards 2018 (Edwards et al., 2018)
6	AI as an evolved system performing tasks as cognitive and affective analysis of the learners.	Chatterjee and Bhattacharjee 2020 (Chatterjee & Bhattacharjee, 2020)
7	AI as a system for designing curriculum for a set of outcomes. AI is utilizing the expected competencies and skill set for input mapping the expected outcome, identifying the gap and thus assisting in designing the curriculum.	M. Somasundaram 2020 (Somasundaram et al., 2020)

of learners' data/record, and their learning tendencies. The patterns related to student profile building and learning growth will be easier to keep track of AI. Concluding how to use assessment findings for better learning will be easier with AI, along with the linking of assessment with personalized feedback.

AI-based techniques such as automation of information, planning and scheduling enrich pedagogical and educational environments. The automation of information, planning and scheduling can be managed using AI-based approaches (Garcia-Penalvo, 2008). The sequencing, prioritization and parallelization of activities is a part of the pedagogical scaffold for a well-designed plan that achieves a set of goals. The well-defined deadlines for activity completion and structured duration of the course assure quality of the education process. The scheduling of activities is required in view of time, space and resource availability. E-learning endows students the power to take control of the learning initiatives and how knowledge is presented during instruction (Atolagbe, 2002). This allows the instructor to assess and frame learning designs in a more critical way (Garrido et al., 2008). Figure 7.2 presents connections between AI and educational applications, showing an intermediate, evolved system of educational technologies.

- Instructor-Centred Pedagogy:
  This is a behaviourist approach where the teacher is at the centre or the focus of the learning process. This pedagogy typically relies on methods such as lecturing, rote memorization, and chorus types of question-response formative assessment. This approach costs learners valuable time spent completing lower-order tasks. The positive side of this pedagogy is that effectiveness could be achieved with increased frequency of asking students to explain and elaborate key ideas, reducing the lecturing component. In the adaptation of this approach, AI-enabled programs may be the centre of pedagogy as a counterpart of the instructor, keeping the learners' role as recipient. AI programs will

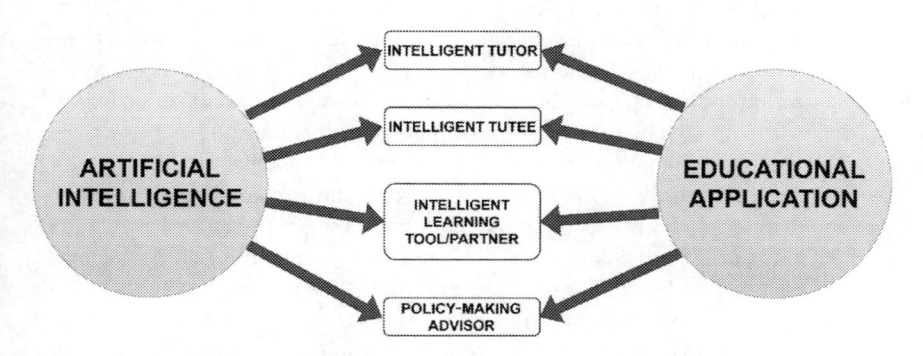

FIGURE 7.2   Newer Roles and Their Interconnection with Pedagogy Equipped with AI (adopted from Gwo-Jen Hwang et al., 2020).

be given some part of the instructors' role and will carry the course content and lead the learning processes. The learner acts as a recipient of the AI and follows a particular predefined learning pathway. The learner acts as a recipient reacting to pre-specified sequences of knowledge, following learning procedures and pathways, and executing learning activities set by AI-enabled programs to achieve predefined goals (Holmes et al., 2019; Koschmann, 1996). AI-enabled programs mimic the teaching machine (Skinner, 1958), create and deliver logical presentations of course content, depending on the learner's clear response presenting correct knowledge spontaneously (Burton et al., 2004). The component of two way exchange and buildup of the learner's newly generated knowledge and skills as a result of learning event and individual response will be missed in this case (Valerie Shute, 1995).

- Learner-Centred Pedagogy:
  This constructivist pedagogical approach is sometimes called inconspicuous as the learning is achieved through participatory, active experiences and reflection, putting the learner at the centre. This approach requires creating learning instances with the learner playing an active role in the learning process. The instructor is a facilitator, and learners build upon prior knowledge and new experiences to create knowledge. In a learner-centric approach in which the needs of the learners are to be catered, it becomes necessary to have a technology-supported platform that collects data about learner's activity, progress rates, performance record, and the time taken at each step. The inspiration of such a pedagogical approach is in keeping the learner at the centre, with AI-enabled programs used as a technological tool supplementing the mentoring role. In this approach a synergetic collaboration among the stakeholders of evolved pedagogy such as the learner, the instructor, and technology, are established. AI-enabled programs work as assistants, providing transparent, accurate and effective environments for learners and instructors to achieve the outcomes of teaching-learning (Riedl, 2019; Yang et al., 2021). The progress of AI-enabled programs reflects the prime objective; that of augmenting human pedagogical skills (Tegmark, 2017).

- Instructor-guided, learner-centred Pedagogy:
  This social constructivism pedagogy puts the learning at the centre, establishing teacher/learner collaborative process. This pedagogy blends the learner-centred and teacher-centred pedagogy, requiring the instructor's flexible attitude in utilizing the local context and environment, and in ensuring learner involvement, as well as availability of teaching and learning aids. This pedagogical approach has the philosophical basis that interaction with people, information and technology in social surroundings leads to learning (Liu & Matthews, 2005). The incorporation of AI-enabled programs and the learner interactions are designed to be mutually active, optimizing learner-centric, personalized learning. The AI-enabled programs are fed with learners' up-to-date individualized information with an aim to optimize the learner model, and the learner has to communicate collaboratively with AI-enabled programs in order to achieve higher and efficient learning outcomes (Boulay, 2019; Toby Baker et al., 2019).

The teaching-learning scaffold aims at creating an understanding about how learners learn and master content. To achieve this aim, it must identify and create the learning environments. The pedagogy of education revolves around raising educators' awareness and demystifying educational phenomena. The educational set-up generates enormous data from various sources, with different formats and availabilities. It is the enormous and complex nature of the data available that demands for sophisticated data mining techniques (C. Romero & Ventura, 2013). The data mining helps in utilizing this information at the level of a group of students up to the institutional levels. Educational data mining (EDM) is a computational approach towards detecting patterns in the educational data repositories. It is up to educators and educational institutions to access the help available from the data-driven options for making creative instructional decisions and remodelling instruction (Labarthe et al., 2018; Rienties et al., 2020).

7.3 DESIGN OF PEDAGOGICAL SOLUTIONS BASED ON AI

The AI-based educational tools with effective use of physical/biological changes as a biomarker will be helpful in the learner's affective domain response (Yee Chung et al., 2021). The evaluation of learning with the biomarker of heartrate variability to measure feelings of sorrow and delight will help detect learners' emotions during the learning experience, and thus instructors can make necessary changes in their instructional design. The inclusion of AI in detecting participants' emotions in game-based learning on the basis of ML techniques is one of the focus areas. Through a balance of game play and course content, participants are guided to apply learning in real world situations. Game play strategies set a competitive situation between oneself or between groups and assess the mastery of content with a reward/penalty system. Fluctuations in the facial expressions related to the learnings gained in a game-based event are reported by Manuel Ninaus et al. (Ninaus et al., 2019).

A learning target achieved through a game-based learning event is compared with a non-game-based equivalent through automated detection of emotions supported with the rankings provided by participants. Another study reported by Lester et al. involves a game-based learning approach in which the gaming technology was blended in AI frameworks (Lester et al., 2013). The design was done with AI-based smart tutoring narrative technologies to create personalized learning experiences. The effective outcomes were achieved by relying on the capabilities of this approach, which included portrait-based dynamic scenarios, concurrent evaluation, and intelligent game-based learning settings. The interconnection between AI and serious gaming has been reportedly established by a programme to play chess (Westera et al., 2020). AI domains such as computational intelligence, algorithms, and machine learning have developed in the direction of human expertise; competition in rule-based strategy games such as Chess and Poker (Fujita & Wu, 2012). AI methods have contributed towards enhancing graphical realism, setting levels, sceneries and storylines, player profile building and moderating complexities via allocating intelligent behaviours to fictitious characters (Yannakakis & Togelius, 2015).

The game play-based pedagogies would need contributions from AI mainly in three sections: 1. Personalized learning, 2. Task automation, and 3. Support outside the classroom. The personalized learning that AI can add in game play-based peda-gogies is enabling adaptive learning by atomization and constant modifications in the learning content as per the learner's skill and understanding. The accumulated data on a player's actions, abilities and learning styles would be analyzed and processed further to synthesize the content designed specifically to cater personalized learn-ing requirements. This would ultimately evolve as a situation wherein the learner is moving hastily through content material and wishes for a greater challenge—or con-versely, the deployment of extra aid mechanisms when a learner is unable to grasp a particular concept. The functionality of automation could be a boon to AI-enhanced learning games; by assigning allied tasks of instructions to AI tools embedded within learning games, the instructors will be spared time, giving them more availability to engage directly with learners providing personalized attention and additional sup-port. The AI-enabled educational games will serve as powerful tools for supporting learners outside the classroom. There is a significant hurdle in providing learning support outside the classroom to students who face socioeconomic barriers and have limited educational support available at home. The learning activities conducted through games enhanced with AI can tackle this by providing tailored support to players directly within the game, helping to match the playing conditions and ensur-ing the support required is made available, irrespective of the learning environment.

Predicting learner performance, gaining insight into student retention, assessing learner requirements and providing course content as per their capabilities are a few of the prominent front areas where pedagogy scientists have sought help from AI. Accurate prediction about learners' growth and performance based on their his-tory and current academic records is crucial for effectively conducting necessary pedagogical interventions to ensure students' graduations up to the mark and in the defined time framework. The focus points in this direction are: differences in socio-economic backgrounds, variations in course selection tendencies, differences in skill sets and knowledge levels, and the need to incorporate personalized learners' learn-ing tendencies and evolving progress. A system of multiple base predictors along with the data-driven approach considering hidden factors and probabilistic matrix factorization for efficacious predictions was designed by Jie Xu et al. (Xu et al., 2017). It was claimed to address the focus points with the development of machine learning methods for predicting student performance.

The student retention prophecy could be effectively done using ML techniques. A retention prediction model using ML algorithms with developed statistics is avail-able in the literature. An accuracy of 97% was achieved in predictions of student retention on the basis of the first three weeks of the course. The results thus obtained were fitted in the pedagogical context to extend the support to the learners, ensuring higher retention and better learning. The ML part of the work comprised Nearest Neighbour (1-NN) classifier using five categorical variables. The Sequential Forward Selection (SFS) was used by selecting the highest accuracy individual feature fol-lowed by its pairing with all others to find the strongest pair.

The overlapping areas of AI and educational data mining can be influential in the teaching-learning process. The three independent fields 1) pedagogies equipped with the ease of computational facilities, 2) support of learning analytics and 3) smart machines mimicking natural intelligence can overlap as shown in Figure 7.3.

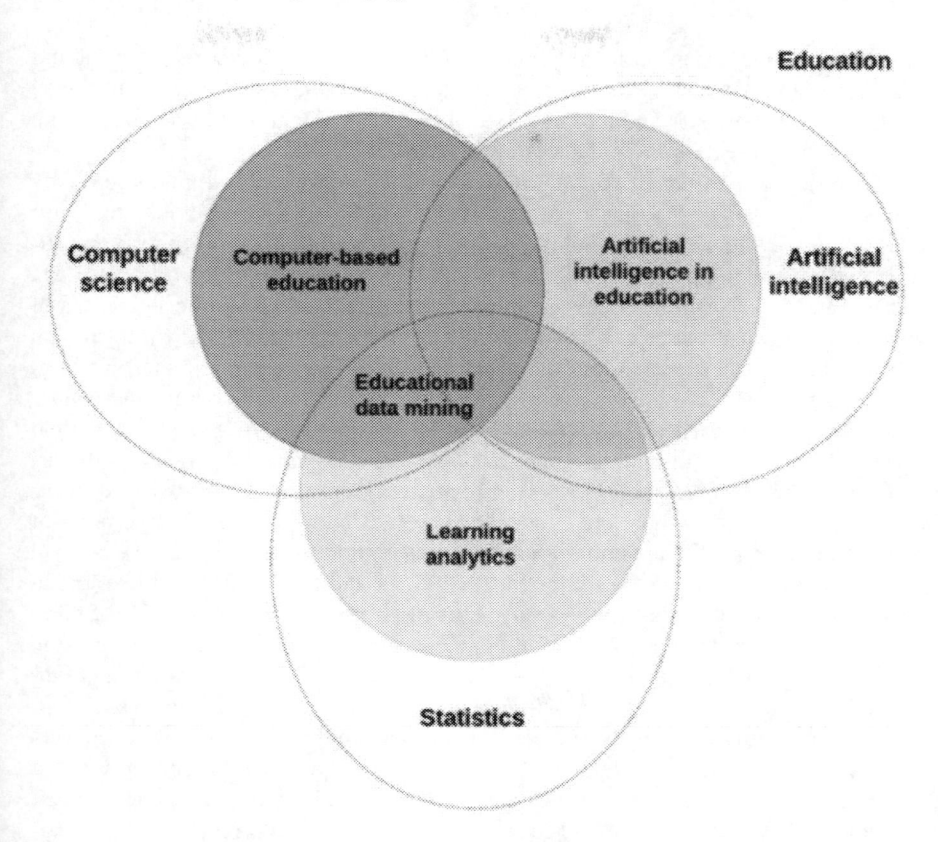

FIGURE 7.3 The Fields of Computer Science, Learning Analytics and Artificial Intelligence Overlap for Evolved Pedagogies (Chen et al., 2020; M. Romero et al., 2017).

The educational institutions have already started embracing such approaches integrating the AI component in the Educational Data Mining (EDM) to "track" students' behaviours, figuring out losses such as students' dropping out, so that it will offer well-timed assists via the regular evaluation of facts and track records in the class, and timely compliance of assignments (Luckin et al., 2016). The EDM designs have evolved as predictive models with data mining platforms. EDM helps in better comprehension of the dataset and the context of collected data. A method involving machine learning algorithms enabling the analysis of educational data in an iterative process where the knowledge discovery and the accuracy of the predictive model is studied by Toivonen et al. (Toivonen et al., 2019). An increase in accuracy with time and possible knowledge discovery via the accumulated clustered data processed with Neural N-Tree models is reported in this article. The early alert mechanism is a smart intervention that educational institutes can incorporate for timely identification of at-risk learners (Atif et al., 2020). The significance of student early alert systems is that support could be offered to high-risk students while they are still enrolled in the unit and able to influence their learning achievements before the completion of the unit. The requirement from the instructor is only identifying the at-risk student and designing further teaching/learning activities in a way that would bring change in the learners' behaviours and learning tendencies. The system will keep sending the formal timely interventions and alerts to the learners.

The rationale behind increasing interest in adopting ML algorithms is that they create possibilities for adaptive learning systems. The amalgamation of item response theory and ML algorithms resulted in an innovative system for dealing with the difficulty of supplying content to new users of adaptive learning systems (Pliakos et al., 2019). The ML algorithm-based system was able to predict the appropriate levels of content for new users. The consent of the learner participant to let their information be collected and processed is an important as well as dynamic parameter in EDM. The consent provided by the learners varies over socioeconomic, ethnic and gender bases. The use of learners' educational data for learning analytics raises an ethical dilemma around how to maintain privacy about learners' data collection and its further usage. Furthermore, there is a possibility that the learners' feedback in the learning analytics may skew predictive models, as well as unintentional incorporation of biases and disproportionate opting out of predictive models (Li et al., 2021).

The dream of ensuring accessible opportunities and education to all learners at all levels includes students with various learning differences (Tucker, 2016; Shipra Sharma & Shalini Garg, 2020). AI-equipped pedagogies have revealed promising results to help differently abled learners, as in the case of communication and language learning (Popenici & Kerr, 2017). The inclusion of wearable devices is helping learners facing challenges in audio-visual impairments. A recent report covers AI-powered diagnostic tools for automated detection of special needs such as reading disabilities, mathematical disabilities and linguistic disabilities or Attention Deficit Hyperactivity Disorder (ADHD) (Drigas & Ioannidou, 2013). Tendencies and patterns such as short attention spans and teasing/bullying from peers have negative impacts on their learning. A learner with such disabilities may suffer socially and be left behind. AI can help convert the difficult content to alternatives more easily understood (Lynch, 2018). AI-enabled pedagogies could take the form of special tools and lessons specifically designed and developed for learners with special needs.

The use of AI as a supportive alternative to the instructor is a growing trend. Studies have been conducted related to teacher response tools and Pedagogical Conversational Agents (Atif et al., 2021; Bywater et al., 2019). The common problem instructors face is the struggle to give an immediate and effective response to learners' mathematical questions, as the time window to efficaciously analyze the learners' perspective is limited. This task may become pretty tough for instructors as they are trying to comply with multiple, competing deadlines and targets. This has led to the introduction of teacher response tools with the use of NLP techniques. The tool backs the instructor, noticing gists of the students' course-related thoughts and supplies quality responses to learners' thoughts. Moreover, it also supports automated and powerful reply suggestions to teachers without simplifying the complexity of the material. Secondly, the use of AI as a Pedagogical Conversational Agent (CAs) is also materialized and practiced. The role of these agents is to share the expertise and knowledge gained, as if with a teacher or peer learner. The additional benefit with these agents is anonymous access at any time, there is no boredom at the agent's end about the same query asked repeatedly and their ability to let the learners explore choices, scenarios and consequences.

7.4 CONCERNS AND POSSIBLE CHALLENGES

Experts have sounded alarms about the startling effects of AI on society in the last two decades (Bostrom, 2014; Kurzweil, 2000, 2006). The prominent concerns about the use of AI in education are: 1) AI could go wrong in the sense of malfunctioning about ethical, social, and humanitarian aspects, 2) technology could replace teachers, and 3) AI could take over and start functioning on its own beyond human control. Whenever a system is implemented, requires certain resources as well as know-how training for the stakeholders. The resources that AI needs beyond the computational facilities are a 'large' volume of 'correct' data. 'Large' indicates exhaustive, rather than all possible data of a given environment. AI can give accurate and authentic results when it is fed with this data (LeCun et al., 2015). 'Correct' data is accurate and covers a substantial portion of a given population, so that when applied to general as well as specific situations, it gives unbiased results without statistical errors (Barocas & Selbst, 2018; Sessions & Marco, 2006). The manual to AI transition phase would pose a difficulty in feeding the records repositories maintained manually on paper. These data records need to be in digitized form to feed to AI. AI programs accelerate data collection, processing, analysis, and interpretation of concurrent data (Wang, 2021). The protection of data is the end-point concern once all the relevant data are available in digital form, accumulated, verified for correctness and fed to the AI program. The risk of breaches on data security could create lifelong negative impacts on stockholders of educational process (Vincent-Lancrin & Vlies, 2020).

Malfunctioning incidents have been reported about the failure of online meeting platforms to prevent posting of obscene content (Ghouri & Ghouri, 2021). Another incident occurred when an algorithm using economic and societal bias generated an outcry over grade prediction (Theodoros et al., 2020). The algorithm downgraded a majority of the students and there was a major disparity between grades predicted by teachers and by the algorithm. This disaster is an example about being careful about AI implementation. It is inherent for AI-based algorithms to amplify trends or patterns as a part of their functioning. One incident reportedly happened with Amazon. com Inc.'s recruitment process (Dastin, 2018). The job of finding the most suitable candidate for a given job profile was conducted through an AI program. Gender biasing was integrated into the programme by observing the existing staff patterns. The AI had learned that most existing staff were male and resultingly disfavored the applications of female job candidates.

The teaching-learning process is 'by human for human'; reducing the individual learner to a mere data point is a great risk. AI programs bestowed with the power to generate decisions will produce analytical, calculation-based results, which in the real world may create an ethical minefield with cold hearted and emotionally detached decisions. The lack of human touch may fail to achieve the learners' well-being and character building, though this is true for any other new technology being introduced into a situation. The incidents caused concern about AI being reliable and ethical in work areas. The important question in this context is whether the concerns about AI are being exaggerated. Lack of trust due to misconceptions about AI systems leads to provisions making AI algorithms safer and more reliable. The education sector needs to embrace AI in pedagogy because its prevalence will become unavoidable, and the delay to include it will most probably refrain our growth. There is no fear about instructors being replaced by technology, but instructors will replace those who will fail to use technology effectively (Farr & Murray, 2016).

7.5 CONCLUSION

The critique presented in this chapter brings out the important points as follows:

1. Pedagogical approaches will get modified with the inclusion of AI technologies such as visual recognition, speech recognition, expert systems and natural language processing.
2. The use of AI enabled ICT tools or devices such as mobile devices, wearable devices and robots is becoming a major medium of instructions in the present-day post-Covid-19 situation. Such systems are modifying pedagogies across all domains of education, such as arts and social sciences, STEM courses, the medical field and more.
3. The key to effective and fruitful use of AI education technologies is in how instructors keep up with the fast-paced developments and include them in learning models or implementation frameworks by thorough consideration about compatibilities of emerging technologies and educational theories.
4. The benefits that the inclusion of AI in educational pedagogies could bring are increased flexibility, personalization and effectiveness in the teaching-learning process.

REFERENCES

Ascione, L. (2017). Brace yourselves: AI is set to explode in the next 4 years. *ESchool News*. Available at: www.eschoolnews.com/2017/05/22/brace-ai-set-explode-next-4-years/.

Atif, A., Jha, M., Richards, D., & Bilgin, A. A. (2021). Chapter 1—Artificial Intelligence (AI)-enabled remote learning and teaching using pedagogical conversational agents and learning analytics. In S. Caballé, S. N. Demetriadis, E. Gómez-Sánchez, P. M. Papadopoulos, & A. Weinberger (Eds.), *Intelligent systems and learning data analytics in online education* (pp. 3–29). Academic Press. https://doi.org/10.1016/B978-0-12-823410-5.00013-9.

Atif, A., Richards, D., Liu, D., & Bilgin, A. A. (2020). Perceived benefits and barriers of a prototype early alert system to detect engagement and support 'at-risk' students: The teacher perspective. *Computers & Education*, *156*, 103954. https://doi.org/10.1016/j.compedu.2020.103954.

Atolagbe, T. A. (2002). *E-learning: The use of components technologies and Artificial Intelligence for management and delivery of instruction*. In ITI 2002. Proceedings of the 24th International Conference on Information Technology Interfaces (IEEE Cat. No.02EX534), vol. 1, pp. 121–128. https://doi.org/10.1109/ITI.2002.1024662.

Baker, T., Smith, L., & Anissa, N. (2019). *Educ-AI-tion rebooted? Exploring the future of Artificial Intelligence in schools and colleges*. Nesta. Available at: www.nesta.org.uk/report/education-rebooted/.

Barocas, S., & Selbst, A. D. (2018). Data quality and Artificial Intelligence—mitigating bias and error to protect fundamental rights. *European Union Agency for Fundamental Rights*, *20*.

Bostrom, N. (2014). *Superintelligence: Paths, dangers, strategies*. Oxford: Oxford University Press.

Boulay, B. du. (2019). Escape from the skinner box: The case for contemporary intelligent learning environments. *British Journal of Educational Technology*, *50*(6), 2902–2919. https://doi.org/10.1111/bjet.12860.

Broggy, J., O'reilly, J., & Erduran, S. (2017). Interdisciplinarity and science education. In K. S. Taber & B. Akpan (Eds.), *Science education: An international course companion* (pp. 81–90). SensePublishers. https://doi.org/10.1007/978-94-6300-749-8_6.

Burton, J. K., Moore, D. M. (M.), & Magliaro, S. G. (2004). Behaviorism and instructional technology. In D. H. Jonassen (Ed.), *Handbook of research on educational communications and technology* (pp. 3–36). Mahwah, NJ: Lawrence Erlbaum Associates Publishers.

Bywater, J. P., Chiu, J. L., Hong, J., & Sankaranarayanan, V. (2019). The teacher responding tool: Scaffolding the teacher practice of responding to student ideas in mathematics classrooms. *Computers & Education*, *139*, 16–30. https://doi.org/10.1016/j.compedu.2019.05.004.

www.carnegielearning.com. (2021). *K-12 education solutions provider*. Carnegie Learning. Available at: www.carnegielearning.com.

Chatterjee, S., & Bhattacharjee, K. K. (2020). Adoption of Artificial Intelligence in higher education: A quantitative analysis using structural equation modelling. *Education and Information Technologies*, *25*(5), 3443–3463. https://doi.org/10.1007/s10639-020-10159-7.

Chen, X., Xie, H., Zou, D., & Hwang, G.-J. (2020). Application and theory gaps during the rise of Artificial Intelligence in education. *Computers and Education: Artificial Intelligence*, *1*, 100002.

www.cognilytica.com/. (2018). AI today podcast #052: AI on the campus—interview with John Rome, Arizona State University. *Cognilytica*. Available at: www.cognilytica.com/.

Dastin, J. (2018). Amazon scraps secret AI recruiting tool that showed bias against women l *Reuters*. Available at: www.reuters.com/article/us-amazon-com-automation-insight/amazon-scraps-secret-ai-recruiting-tool-that-showed-bias-against-women-idUSKCN1MK08G.

Drigas, A., & Ioannidou, R.-E. (2013). A review on Artificial Intelligence in special education. In *Communications in computer and information science* (Vol. 278). https://doi.org/10.1007/978-3-642-35879-1_46.

du Boulay, B. (2016). Artificial Intelligence as an effective classroom assistant. *IEEE Intelligent Systems*, *31*(6), 76–81. https://doi.org/10.1109/MIS.2016.93.

Duong, M. T., Rauschecker, A. M., Rudie, J. D., Chen, P.-H., Cook, T. S., Bryan, R. N., & Mohan, S. (2019). Artificial Intelligence for precision education in radiology. *The British Journal of Radiology*, *92*(1103), 20190389. https://doi.org/10.1259/bjr.20190389.

Edwards, C., Edwards, A., Spence, P. R., & Lin, X. (2018). I, teacher: Using Artificial Intelligence (AI) and social robots in communication and instruction. *Communication Education*, *67*(4), 473–480. https://doi.org/10.1080/03634523.2018.1502459.

Farr, F., & Murray, L. (2016). Introduction: Language learning and technology. In *The Routledge handbook of language learning and technology*. New York: Routledge.

Fujita, H., & Wu, I.-C. (2012). A special issue on Artificial Intelligence in computer games: AICG. *Knowledge-Based Systems*, *34*, 1–2. https://doi.org/10.1016/j.knosys.2012.05.014.

Garcia-Penalvo, F. J. (2008). *Advances in E-learning: Experiences and methodologies*. IGI Global. https://services.igi-global.com/resolvedoi/resolve.aspx?doi=10.4018/978-1-59904-756-0. Available at: www.igi-global.com/book/advances-learning-experiences-methodologies/www.igi-global.com/book/advances-learning-experiences-methodologies/40.

Garrido, A., Onaindia, E., & Sapena, O. (2008). Planning and scheduling in an e-learning environment. A constraint-programming-based approach. *Engineering Applications of Artificial Intelligence*, *21*(5), 733–743. https://doi.org/10.1016/j.engappai.2008.03.009.

Ghouri, H., & Ghouri, A. (2021). Technology enhanced learning: Should Artificial Intelligence ever be used for teaching and learning? *Journal of Educational Sciences & Research, 7*(2), 163–183.

Holmes, W., Bialik, M., & Fadel, C. (2019). *Artificial Intelligence in education: Promises and implications for teaching and learning*. Center for Curriculum Redesign. Available at: http://udaeducation.com/wp-content/uploads/2019/05/Artificial-Intelligence-in-Education.-Promise-and-Implications-for-Teaching-and-Learning.pdf.

www.holoniq.com/. (2021). *$16.1B of global EdTech venture capital in 2020*. HolonIQ. Available at: www.holoniq.com/notes/16.1b-of-global-edtech-venture-capital-in-2020/.

Hwang, G.-J. (2003). A conceptual map model for developing intelligent tutoring systems. *Computers & Education, 40*(3), 217–235. https://doi.org/10.1016/S0360-1315(02)00121-5.

Hwang, G.-J., Xie, H., Wah, B. W., & Gašević, D. (2020). Vision, challenges, roles and research issues of Artificial Intelligence in education. *Computers and Education: Artificial Intelligence, 1*, 100001. https://doi.org/10.1016/j.caeai.2020.100001.

Johnson, B. G., Phillips, F., & Chase, L. G. (2009). An intelligent tutoring system for the accounting cycle: Enhancing textbook homework with Artificial Intelligence. *Journal of Accounting Education, 27*(1), 30–39. https://doi.org/10.1016/j.jaccedu.2009.05.001.

www.knewton.com/. (2021). *Knewton—Achievement within reach*. Knewton. Available at: www.knewton.com/.

Koschmann, T. (1996). *Paradigm shifts and instructional technology* (p. 28). https://doi.org/10.1.1.454.2158&rep=rep1&type=pdf.

Kuch, D., Kearnes, M., & Gulson, K. (2020). The promise of precision: Datafication in medicine, agriculture and education. *Policy Studies, 41*(5), 527–546. https://doi.org/10.1080/01442872.2020.1724384.

Kurzweil, R. (2000). *The age of spiritual machines: When computers exceed human intelligence*. London: Penguin Books.

Kurzweil, R. (2006). *The singularity is near by Ray Kurzweil: 9780143037880 | PenguinRandomHouse.com: Books*. Penguin Books. Available at: www.penguinrandomhouse.com/books/291221/the-singularity-is-near-by-ray-kurzweil/.

Labarthe, H., Luengo, V., & Bouchet, F. (2018). *Analyzing the relationships between learning analytics, educational data mining and AI for education. 14th International Conference on Intelligent Tutoring Systems (ITS): Workshop Learning Analytics, 10*.

LeCun, Y., Bengio, Y., & Hinton, G. (2015). Deep learning | Nature. *Nature Review Articles, 521*, 436–444. https://doi.org/10.1038/nature14539.

Lester, J. C., Ha, E. Y., Lee, S. Y., Mott, B. W., Rowe, J. P., & Sabourin, J. L. (2013). Serious games get smart: Intelligent game-based learning environments. *AI Magazine, 34*(4), 31–45. https://doi.org/10.1609/aimag.v34i4.2488.

Li, W., Sun, K., Schaub, F., & Brooks, C. (2021). Disparities in students' propensity to consent to learning analytics. *International Journal of Artificial Intelligence in Education*. https://doi.org/10.1007/s40593-021-00254-2.

Lin, H.-C., Tu, Y.-F., Hwang, G.-J., & Huang, H. (2021). From precision education to precision medicine: Factors affecting medical staff s intention to learn to use AI applications in hospitals. *Educational Technology & Society, 24*(1), 123–137.

Liu, C., & Matthews, R. (2005). Vygotsky's philosophy: Constructivism and its criticisms examined. *International Education Journal, 6*, 386–399.

Luckin, R., Holmes, W., Griffiths, M., Corcier, L. B., Pearson (Firm), & University College, L. (2016). *Intelligence unleashed: An argument for AI in education*. Available at: www.pearson.com/content/dam/corporate/global/pearson-dot-com/files/innovation/Intelligence-Unleashed-Publication.pdf.

Lynch, M. (2018). Using Artificial Intelligence to help students with learning disabilities learn. *The Tech Edvocate*. Available at: www.thetechedvocate.org/using-artificial-intelligence-help-students-learning-disabilities-learn/.

www.netexlearning.com/en/. (2019, January 11). Netex—Learning technologies. *Netex*. Available at: www.netexlearning.com/en/.

Ninaus, M., Greipl, S., Kiili, K., Lindstedt, A., Huber, S., Klein, E., Karnath, H.-O., & Moeller, K. (2019). Increased emotional engagement in game-based learning—A machine learning approach on facial emotion detection data. *Computers & Education, 142,* 103641. https://doi.org/10.1016/j.compedu.2019.103641.

Pliakos, K., Joo, S.-H., Park, J. Y., Cornillie, F., Vens, C., & Van den Noortgate, W. (2019). Integrating machine learning into item response theory for addressing the cold start problem in adaptive learning systems. *Computers & Education, 137,* 91–103. https://doi.org/10.1016/j.compedu.2019.04.009.

Popenici, S. A. D., & Kerr, S. (2017). Exploring the impact of Artificial Intelligence on teaching and learning in higher education. *Research and Practice in Technology Enhanced Learning, 12*(1), 22. https://doi.org/10.1186/s41039-017-0062-8.

Riedl, M. O. (2019). Human-centered Artificial Intelligence and machine learning. *Human Behavior and Emerging Technologies, 1*(1), 33–36. https://doi.org/10.1002/hbe2.117.

Rienties, B., Køhler Simonsen, H., & Herodotou, C. (2020). Defining the boundaries between Artificial Intelligence in education, computer-supported collaborative learning, educational data mining, and learning analytics: A need for coherence. *Frontiers in Education*. https://doi.org/10.3389/feduc.2020.00128.

Rita Liao. (2018). Tencent-backed homework app jumps to $3B valuation after raising $300M. *TechCrunch*. Available at: https://social.techcrunch.com/2018/12/26/yuanfudao-raises-300-million/.

Romero, C., & Ventura, S. (2013). Data mining in education. *WIREs Data Mining and Knowledge Discovery, 3*(1), 12–27. https://doi.org/10.1002/widm.1075.

Romero, M., Lepage, A., & Lille, B. (2017). Computational thinking development through creative programming in higher education. *International Journal of Educational Technology in Higher Education, 14*(1), 1–15.

Sessions, V., & Marco, V. (2006). The effects of data quality on machine learning algorithms. *Mitiq MIT Education Documents, 14.*

Shipra Sharma & Shalini Garg. (2020). Impact of Artificial Intelligence in special need education to promote inclusive pedagogy. *International Journal of Information and Education Technology, 10*(7), 523–527. https://doi.org/10.18178/ijiet.2020.10.7.1418.

Skinner, B. F. (1958). Teaching machines: From the experimental study of learning come devices which arrange optimal conditions for self-instruction. *Science, 128*(3330), 969–977. https://doi.org/10.1126/science.128.3330.969.

Somasundaram, M., Latha, P., & Pandian, S. A. S. (2020). Curriculum design using Artificial Intelligence (AI) back propagation method. *Procedia Computer Science, 172,* 134–138. https://doi.org/10.1016/j.procs.2020.05.020.

Tegmark, M. (2017). *Life 3.0: Being human in the age of Artificial Intelligence*. London: Penguin.

Theodoros, E., Hardoon, D. R., & Ovchinnikov, A. (2020, August 13). What happens when AI is used to set grades? *Harvard Business Review*. Available at: https://hbr.org/2020/08/what-happens-when-ai-is-used-to-set-grades.

Toivonen, T., Jormanainen, I., & Tukiainen, M. (2019). Augmented intelligence in educational data mining. *Smart Learning Environments, 6*(1), 10. https://doi.org/10.1186/s40561-019-0086-1.

Tucker, E. (2016). *Artificial Intelligence and disability: An academic study of AI use in the classroom for students with disabilities*. Available at: www.elijahatucker.me/projects/Tucker_Elijah_Thesis.pdf.

Valerie Shute. (1995). SMART: Student modeling approach for responsive tutoring | SpringerLink. *User Modeling and User-Adapted Interaction, 5*. https://doi.org/10.1007/BF01101800.

Vincent-Lancrin, S., & Vlies, R. van der. (2020). *Trustworthy Artificial Intelligence (AI) in education: Promises and challenges*. https://doi.org/10.1787/a6c90fa9-en.

Vlasova, E. Z., Goncharova, S. V., Barakhsanova, E. A., Karpova, N. A., & Ilina, T. S. (2019). Artificial Intelligence for effective professional training of teachers in the Russian Federation. *Revista ESPACIOS, 40*(22).

Wang, Y. (2021). When Artificial Intelligence meets educational leaders' data-informed decision-making: A cautionary tale. *Studies in Educational Evaluation, 69*, 100872. https://doi.org/10.1016/j.stueduc.2020.100872.

Westera, W., Prada, R., Mascarenhas, S., Santos, P. A., Dias, J., Guimarães, M., Georgiadis, K., Nyamsuren, E., Bahreini, K., Yumak, Z., Christyowidiasmoro, C., Dascalu, M., Gutu-Robu, G., & Ruseti, S. (2020). Artificial Intelligence moving serious gaming: Presenting reusable game AI components. *Education and Information Technologies, 25*(1), 351–380. https://doi.org/10.1007/s10639-019-09968-2.

Woolf, B. (1988). Intelligent tutoring systems: A survey. In *Exploring Artificial Intelligence* (pp. 1–43). San Francisco, CA: Morgan Kaufmann Publishers Inc.

Xu, J., Moon, K. H., & van der Schaar, M. (2017). A machine learning approach for tracking and predicting student performance in degree programs. *IEEE Journal of Selected Topics in Signal Processing, 11*(5), 742–753. https://doi.org/10.1109/JSTSP.2017.2692560.

Yang, S. J. H. (2021). Guest editorial: Precision education—A new challenge for AI in education. *Educational Technology & Society, 24*(1), 105–108.

Yang, S. J. H., Ogata, H., Matsui, T., & Chen, N.-S. (2021). Human-centered Artificial Intelligence in education: Seeing the invisible through the visible. *Computers and Education: Artificial Intelligence, 2*, 100008. https://doi.org/10.1016/j.caeai.2021.100008.

Yannakakis, G. N., & Togelius, J. (2015). A panorama of artificial and computational intelligence in games. *IEEE Transactions on Computational Intelligence and AI in Games, 7*(4), 317–335. https://doi.org/10.1109/TCIAIG.2014.2339221.

Yee Chung, J. W., Fuk So, H. C., Tak Choi, M. M., Man Yan, V. C., & Shing Wong, T. K. (2021). Artificial Intelligence in education: Using heart rate variability (HRV) as a biomarker to assess emotions objectively. *Computers and Education: Artificial Intelligence, 2*, 100011. https://doi.org/10.1016/j.caeai.2021.100011.

Zhang, K., & Aslan, A. B. (2021). AI technologies for education: Recent research & future directions. *Computers and Education: Artificial Intelligence, 2*, 100025. https://doi.org/10.1016/j.caeai.2021.100025.

Zhuang, Y., Cai, M., Li, X., Luo, X., Yang, Q., & Wu, F. (2020). The next breakthroughs of Artificial Intelligence: The interdisciplinary nature of AI. *Engineering, 6*(3), 245–247. https://doi.org/10.1016/j.eng.2020.01.009.

8 Artificial Intelligence in Assessment of Students' Performance

Suvojit Dhara, Sheshadri Chatterjee, Ranjan Chaudhuri, Adrijit Goswami and Soumya Kanti Ghosh

CONTENTS

8.1 INTRODUCTION

In the 21st century, when the world is moving at a speed beyond imagination, everything needs a massive update to be in the race. From tech industries to the business world, the health sector to military organizations, everything is getting updated with the new-coming technologies, often profiting like never before. Artificial Intelligence (AI) is one such tool helping everything to advance faster than ever before. We are edging towards a world where robots will play an important role with hand in hand with humans. For this advancement, Artificial Intelligence can be thought of as a "Brahmastra" as in the "Hindu Shastra". Artificial Intelligence has impacted almost every sector of our world, from technological advancements to business administration, security agencies, navigation, research and more; resulting in massive improvement and growth (Chatterjee et al., 2021). These are all profitable fields. Various non-profit sectors like health or education are also not left behind. Various studies have highlighted how AI has helped to improve the infrastructure of the health (Jiang, et al., 2017; Davenport & Kalakota, 2019) and the education sectors (Subrahmanyam & Swathi, 2018; Chen et al., 2020). In this chapter, we will study such an application of AI in the education field, in particular in assessing students' performance and improvement.

8.1.1 ARTIFICIAL INTELLIGENCE IN EDUCATION

Education is the backbone of a society, and we may consider the global human race as nothing but a bigger society. When we talk about the development or growth of the human race in every field, education cannot be left behind. Data Mining (DM), the main branch or tool in AI, assesses examples and finds useful patterns, which are in turn processed for improvement of the concerned system. Similarly, data mining helps educational institutions to assess previous records in their student database, and even merge databases with other stakeholders to make required changes towards improvement (Chatterjee & Bhattacharya, 2020). This branch of research is called 'Educational Data Mining' (EDM). Under the umbrella of EDM, the task of student performance assessment (Shahiri et al., 2015) seems most important; research works are underway to discover the possible relationships between various social, economic and environmental factors and academic performance. Additionally, various AI tools such as recommendation systems (Yadav et al., 2016) and educational gaming apps (Zirawaga et al., 2017) help the learner in a much better scope, and make education easy and interesting for all. We will discuss more about this in the next section. Additionally, various other important tasks are conducted in the field of EDM; for instance, student profiling, where students are clustered according to their various hard skills (academic backgrounds, grades, achievements, etc.) and soft skills (communication, behavior, attitude, hobbies, etc.) using DM techniques (Bouchet et al., 2013), enrolment management where various effective student enrolment strategies are made with the use of data from previous enrolment records (Aulck et al., 2020), syllabus organization where research is conducted towards forming inter-related, effective and student-friendly curricula in order to upgrade student performance, and more. However, in this chapter we shall constrain ourselves to discussing the role of AI in the task of assessing and improving students' performance.

The never-ending advantages of AI have drawn the attention of education researchers and motivated them to incorporate various tools of AI into the education of today's world so that massive improvements can be achieved in terms of not only students' performance but also to various other dimensions of educational settings as well. According to Verified Market Research, the Global Artificial Intelligence in Education Market was valued at $521.03 million in 2018 and is projected to reach $1,0381.70 million by 2026, growing at a CAGR of 45.12%

from 2019 to 2026 (Artificial Intelligence in Education Market Size and Forecast, 2020). In fact, recently UNESCO has published its first-ever document offering guidance and recommendations on how AI technologies can be harnessed best to achieve excellence in Education (cited from https://en.unesco.org/news/first-ever-consensus-artificial-intelligence-and-education-published-unesco).

UNESCO has published the Beijing Consensus on Artificial Intelligence (AI) and Education, the first-ever document to offer guidance and recommendations on how best to harness AI technologies for achieving the Education 2030 Agenda. It was adopted during the International Conference on Artificial Intelligence and Education, held in Beijing from 16–18 May 2019, by over 50 government ministers, international representatives from over 105 Member States, and almost 100 representatives from UN agencies, academic institutions, civil society and the private sector.

The Consensus states that the systematic integration of AI in education has the potential to address some of the biggest challenges in education today, innovate teaching and learning practices and ultimately accelerate the progress towards SDG 4.

In summary, the Beijing Consensus recommends governments and other stakeholders in UNESCO's Member States to:

1. *Plan AI in education policies in response to the opportunities and challenges AI technologies bring, from a whole-government, multi-stakeholder, and inter-sectoral approach, that also allow for setting up local strategic priorities to achieve SDG 4 targets*
2. *Support the development of new models enabled by AI technologies for delivering education and training where the benefits clearly outweigh the risks, and use AI tools to offer lifelong learning systems which enable personalized learning anytime, anywhere, for anyone*
3. *Consider the use of relevant data where appropriate to drive the development of evidence-based policy planning*
4. *Ensure AI technologies are used to empower teachers rather than replace them, and develop appropriate capacity-building programs for teachers to work alongside AI systems*
5. *Prepare the next generation of existing workforce with the values and skills for life and work most relevant in the AI era*
6. *Promote equitable and inclusive use of AI irrespective of disability, social or economic status, ethnic or cultural background or geographical location, with a strong emphasis on gender equality, as well as ensure ethical, transparent, and auditable uses of educational data.*

The goal of this chapter is to attract the readers' attention readers to the current research directions adapted in the field of EDM and moreover AI in the education sector to upgrade students' performance levels, and also to the development of their overall personality, to some extent. The rest of the chapter is arranged as follows. In section two, we briefly discuss 'Students' Performance Assessment' and the role of AI in this regard. In section three, we will discuss various factors influencing a student's performance. In section four, we will discuss how AI is deployed in various online learning platforms to assess the influence of various factors on students' performance (Chatterjee, Ghosh et al., 2020). Finally, in section five, we will try to show the direction of existing and future research on how AI can help students in improving their performance. The chapter ends with a conclusion presented in section six.

8.2 STUDENTS' PERFORMANCE ASSESSMENT

Students are said to be the future of a nation. The development of any country depends highly on the students of that country. Hence Educational Data Mining is mostly concerned with students. EDM deals with many problems in the field of education, as discussed earlier, but "students' performance assessment and prediction" is the most important of all. Before going into a deep discussion of how EDM works in this direction, we should first build up a strong idea about what we really mean by 'students' performance'.

8.2.1 STUDENTS' PERFORMANCE

When we talk about students' performance, we usually mean their academic performance, which includes their performance in a certain course they are enrolled in or a certain class of students they are part of. But we cannot keep ourselves limited to only the academic performance of the students, as that is not the only thing that society depends on. It may be that an academically sound student is of no use to society, whereas an academically not-so-well student serves his or her society much better. It's solely because of the personality the student holds—his attitude, body language, and all (Chatterjee, Rana et al., 2020). Thus, the personality of a student can be thought of as an important factor we should look for in a student, and hence, the performance of a student towards her or his personality development should also be taken into consideration. In this chapter, though we focus mainly on academic performance and how that can be assessed and improved using AI technologies, we shall also discuss a few aspects of how a student's personality can be built through AI.

8.2.2 NEED FOR STUDENTS' PERFORMANCE ASSESSMENT

In the pre-Artificial Intelligence era, we used to assess a student's academic performance depending on only a few factors of his or her academic background, such as past scores. But with the help of various AI techniques and research, we could develop the idea that apart from these factors, various social, demographic, behavioral and non-behavioral features also play an important role in his performance. We can now pre-determine a picture of how the students may perform in upcoming exams or various competitions. This helps an institution to categorically identify its students and take necessary steps to make improvements. If we can predict a student's performance in advance, we can identify those students who are most likely to perform badly or even fail. Thus, we can nurture them specially and work towards their performance improvement. Also, we are able to identify the most promising students and give them much better exposure so that they perform more brilliantly. In upcoming sections, we will discuss various factors upon which what a student's performance relies.

8.3 IDENTIFICATION OF FACTORS INFLUENCING STUDENTS' PERFORMANCE

In the pre-AI era, we had some idea that apart from the academic features of a student, various other features relating to his or her social life, demography, surrounding environment and so on may play an important role in his or her performance. But as these factors are not easily measurable, it was impossible to extract their relations formally. With the advancement of technology, we are now able to apply various important AI techniques into the education field to extract these important relations. Also, various factors that we couldn't think of at that time came to light as important causes of a student's high or low performance, with the help of AI technologies.

For the past decade, scientists have been conducting research to discover the hidden relationships between various factors (aside from academic ones) with students' academic performance. When we look deep into the literature of Indian studies on students' performance assessment, we find many researchers looking for various aspects in an educational or non-educational setup affecting a student's performance. Two renowned professors, B.K. Baradwaj and S. Pal, from Purvanchal University, Uttar Pradesh, conducted research on the same students at the university in 2011 (Baradwaj & Pal, 2011). They explored various academic factors such as "Previous semester marks", "Class-test grades", "Seminar performance", "Assignments", "General proficiency", "Attendance", "Lab work" and "End semester marks" and extracted the relationship these factors reveal with a student's performance. With the database of previous years' students, they were able to predict current students' performance efficiently. In the same year, two professors, R. R. Kabra and R.S. Bichkar from S.G.R College of Engineering and Management, Maharashtra, conducted research (Kabra & Bichkar, 2011) on students at the same college and discovered that many personal factors like "Mother's Occupation" and "Father's Occupation" also play an important role in a student's academic performance. Their model achieved success in identifying students who were likely to fail, and their work was used to improve performance. There are many significant research works, both in our nation and abroad, exploring various other factors. We cluster all these features in various groups such as "Academic Background features", "Demographic features", and "Behavioral features". We discuss the factors in more detail to follow.

8.3.1 Academic Background Features

There are various academic background features such as "Previous semester marks", "Class-test grades", "Choice of subject", and "Number of absence days" which play an important role in a student's performance. For example, a student with good marks in a previous semester often performs well in subsequent semesters. Similarly, a student who takes tests seriously during class hours tends to perform well in the finals too. Choice of subject is also an important feature, as a student who takes up a course according to her or his interest performs well in that course. A science-loving student, if he takes up 'Science' as his course during his schooling, tends to perform better than a student who has only taken up science just based on his or her previous marks or just for the sake of good grades. Lastly, a student with more regular attendance in the class tends to understand the subject better than irregular students, and in turn, that helps him or her in perform well in the exams. In other words, class teaching plays just a part in students' performance.

8.3.2 Demographic Features

Various studies in the field of EDM show that apart from the academic features, there are many demographic attributes too that play a significant role in a student's performance; for example, "A Student's surrounding environment", "Involvement of his parents in his studies" and "Gender of the student". We see a pattern where students who come from a peaceful home tend to be perform better in their studies. There are exceptions too, but it is inevitable that society and family or household conditions impact a student in their studies. If we look deeper, this feature can be broken into more sub-parts such as "Number of family members", "Number of rooms in the house", "Availability of separate study-rooms", and so on. Parental intervention in their son/daughter's studies is of much importance. This can have good and bad effects. If parents are well educated, then they can help their child in studies as well. Hence, we see "Parents' Academic Qualification" is also an important aspect. But too much intervention sometimes puts a lot of pressure on the student resulting in degradation of the student's performance. Though the "gender" of a student is an uncontrollable parameter in the discussion, statistical analysis can find useful patterns based on gender. For example, in an institution where we see male students performing significantly better than female students, it will be essential to support female students to help them improve their results (Chatterjee, Majumdar et al., 2020), for instance, by arranging special doubt-clearing sessions. Apart from these major factors, there are other factors such as "Distance from home to school" and "Availability of proper conveyance" which play a significant role in a student's performance.

8.3.3 Behavioral Features

In the pre-data mining era, we could not easily chart the relationships or degrees of relationships between students' various behavioral patterns and their performance. But now, with the help of various data mining techniques, we can significantly discover the influence those features have on performance measures. By students' behavioral patterns or features, we mean especially the classroom behavior of a student. For example, "Regular attendance" of a student is something that plays an important role in their performances. A regular student in the class absorbs all the lessons taught in the class and thus tends to perform better in the exams than those students who have lower attendance. As seen earlier, it can be thought of as an academic background feature too, but not attending classes regularly is also a malpractice associated with the students' behavioral insights. Similarly, "Classroom performance" and "intractability with instructors during class" are other important factors. Certainly, there are various other factors like "Regular assignment solving" and "Attending to doubt-clearing classes" that affect the behavioral patterns of a student. Regular submission of class assignments helps students not only to better understand the subject with more practice, it builds time management skills as well, which in turn helps the student in various competitive examinations. Similarly, a student may have doubts in the lessons taught in the class, but if he or she attends the doubt-clearing classes, then he or she may get a taste of their peers' understandings of the subject, thus sharpening his or her knowledge. In the same line of thinking, we can identify another important behavioral pattern of students in "Group studying". Studying in a group also enables a student to acquire knowledge of her other mate's understanding of the subject and sharpen her knowledge. Also, in a group study environment, one can share his difficulty with his fellow mate and ask him for help. Although these may seem like philosophical things, we now have empirical knowledge regarding these factors; AI and data mining help us understand the degree of these features' contribution to a student's performance.

There may well be various other factors affecting students' performance, such as classroom attentiveness, their boredom towards a subject, or their attitude towards a certain instructor. In the offline classroom environment, even with the help of DM techniques, these features appear difficult to measure. But in today's world, students benefit from online learning platforms provided to them. In online learning platforms, the aforementioned features can be measured with ease. Also, digging up various data coming from the online setup enables us to be introduced to many more important features that we would not consider in offline settings. In the next section, we discuss those in more detail.

8.4 AI-AIDED ANALYSIS OF STUDENTS' PERFORMANCE IN ONLINE LEARNING PLATFORMS

As discussed in the earlier section, there are several behavioral features that are difficult to assess or analyze in an offline, face-to-face classroom environment. However, massive technological upgrades in recent years have enabled us to provide today's generation of students with various online learning platforms, from online tutoring websites, to websites offering online courses, to student-friendly tutoring applications. In such types of online learning platforms, the assessment of behavioral features has become easier, and all the credit goes to the Artificial Intelligence techniques and research. The main tool has been the clickstream data of the student in the online learning platform. These are also referred to as "Microlevel data" in the language of Educational Data Mining. In this section, we discuss these features and various studies regarding the role of AI in assessment.

8.4.1 STUDENTS' COGNITIVE SKILLS AS A COMPONENT OF KNOWLEDGE

Over the past few decades, there has been considerable work using students' clickstream data coming from online learning platforms to make inferences about how students' performance is related to complex cognitive skills within learning activities. Previously, it was difficult to infer complex cognitive factors, but new data mining methods have made it possible to model and track them over time. In various online learning platforms, hundreds to thousands of students typically generate vast numbers of interactions, ranging from magnitudes of ten thousand to millions of interactions. Automated detectors that identify students' behavioural patterns have been developed and applied to those clickstream data sets to identify the degree to which students transferred their knowledge of scientific inquiry between domains. Moussavi et al. (2016) and Sao Pedro et al. (2014) tried improving the outcomes by driving automated scaffolding aimed at improving students' ability with these skills. There has been works of considerable interest studying students' problem-solving strategies and their impact on performance. For instance, Toth et al. (2014) studied students' problem-solving skills within the *MicroDYN* learning environment and clustered how student strategies developed and shifted over time. Similarly, Bauer et al. (2017) examined problem-solving approaches in the scientific discovery game called '*Foldit*', which tasks its users with identifying various protein structures, a biology research task that is difficult to do in a fully automated fashion. Using visualization to understand the clickstream data generated in the game setup, the authors discovered several common problem-solving strategies and associated them with players' performance. Bauer and colleagues noted that understanding these approaches could provide scaffolding to improve the quality of players' solutions.

8.4.2 METACOGNITIVE AND SELF-REGULATORY LEARNING ABILITY OF STUDENTS

Many studies have been conducted by educational data mining researchers to explore the relationships between metacognitive ability and self-regulatory learning (SRL) ability of students in their performance (Roll & Winne, 2015). These constructs examine the ability of a student to self-regulate the learning process, which is especially relevant in less structured systems like Learning Management Systems (LMS), Massive Online Open Courses (MOOCs) and so on. Research in this direction involves modelling the processes and actions undertaken by students within the learning environment to identify possible scaffolds to encourage learning, which system developers and design engineers may use for building better interfaces and experiences (Roll & Winne, 2015; Aleven et al., 2016). Microlevel clickstream data uniquely provides detailed information on students' temporal and sequential patterns of behaviours, which are based on specific actions students undertake and the system design components students utilize during an online course or, in general, online learning. For instance, Park et al. (2017) explored the development and validation of an effort regulation measure using clickstream data on students' previewing and reviewing of the course materials. Students who increased their efforts to review course materials were more likely to pass rather than students who were irregular in reviewing the study materials. Similarly, Park et al. (2018) developed and validated a time management measure that identifies student ignorance towards the course and their unavailability during the live lectures based on student clickstream data in online courses with periodic deadlines. Students receiving grade 'A' had significantly higher time management skills than 'B'-grade students, who had significantly higher time management skills than 'C', 'D' and 'F'-grade students.

Among other important factors, "Teaching ability of a student" and "Help-seeking behaviour" are also researched by EDM. The teachable agent known as "Betty's Brain" (Biswas et al., 2016; Segedy et al., 2015) offers an example. In Betty's Brain, students are given the task of teaching a computer agent called "Betty" by producing causal maps and models describing various science phenomena. Students' ability to teach Betty is evaluated by a second computer agent, Mr. Davis, who gives Betty quizzes and grades her performance based on how well the student instructed Betty. A student with a better understanding of the subject tends to instruct Betty better than others. Ogan et al. (2015) investigated how help-seeking strategies correlate with learning. Lu and Hsiao (2016) studied how student performance during programming correlated with their help-seeking behaviour within discussion forums and determined that more successful learners read posts in a deeper fashion than less successful learners.

8.4.3 IDENTIFICATION OF ACADEMIC EMOTIONS

Online clickstream data from learning platforms allow us to make inferences about "noncognitive" constructs surrounding engagement, motivation and affect. The most thoroughly studied constructs are *academic emotions*, also referred to as affective states: frustration, confusion, boredom, and engaged concentration (sometimes called *flow*). Various affect detectors have been developed in recent times for various learning environments, including intelligent tutoring systems, puzzle games and first-person simulations (Botelho et al., 2018; DeFalco et al., 2018). The capacity of educational data mining techniques to identify academic emotions affords utilization of affective detectors to provide real-time feedback and interventions to learners. For example, DeFalco et al. (2018) used affective detectors in a military training game to address student frustration as students worked through a combat casualty care skill simulation, TC3Sim, for the US Army. By integrating affective detectors into the game, itself, TC3Sim was able to provide feedback messages to students when frustration was identified, leading to improved student learning from pre-test to post-test.

Apart from microlevel clickstream data, the data coming from the systematic computerized collection of students' writing artifacts during online exams, online forum discussions and so on, have also drawn researchers towards the assessment of students' performance. EDM researchers also refer to such data as "Mesolevel data". As students' learning activities are making a shift from offline, face-to-face classrooms to online tutoring websites or online personalized courses, the corpora of students' writing are becoming more and more accessible day by day. Various online competitive exams like the CAT and SAT are the source of such corpora. Adding to this, written discussions on various online forums and online submitted assignments have made the volume of such data larger, resulting in big data. Along with the microlevel data, such corpora data or Mesolevel data are also processed to discover various student behavioural patterns in online platforms affecting their performances. To serve this purpose, various tools have been developed by the researchers. For instance, the Linguistic Inquiry and Word Count tool (Pennebaker et al., 2015) measures psychological features including confidence, leadership ability, authenticity and emotional tone of students. Other approaches include social network analysis to generate inferences about relational positionings. In the next section, we will have a look at how AI or AI-enabled tools can be deployed to improve students' performance, both in offline classroom environments and online learning setups.

8.5 USE OF AI IN ENHANCING STUDENTS' PERFORMANCE

We have already discussed in the previous sections how Artificial Intelligence has been in use in recent times to assess several factors affecting students' performance both in offline, face-to-face classroom setup and in online learning platforms. Artificial Intelligence is also one such thing that not only assesses performance but also takes charge to improve it. In this section, we will discuss several ways AI tools and techniques have been utilized and will be in use in the future to enhance or improve students' performance. We divide the discussion into two subsections, one concerned with the use of AI in offline setup and the other with the use of AI in online learning systems.

8.5.1 ARTIFICIAL INTELLIGENCE FOR IMPROVEMENT OF STUDENTS' PERFORMANCE IN CLASSROOM SETUP

In previous sections, we have already discussed how we can discover various factors in a face-to-face classroom setup affecting students' performance based on various data mining or AI tools. For example, in a high school environment, various administrative features like "number of students and teachers", "number of sections in a class", "use of internet while teaching", play important roles in students' performance. After assessing these critical factors using data mining tools, the administration can rank the factors according to their importance and act accordingly to improve the features, which in turn will positively affect student's performance. These features are often named as "Macrolevel features" in terms of EDM language. Though this plays an important role in improving the students' performance, this is not as important as the behavioral patterns of a student towards their performance. We have already seen it. Now in the online setup, the behavioral attributes of students can be captured using various tools, but in the offline setup, the task seems difficult. But AI can rescue us in that case too. For instance, the administration can install CCTV cameras in the classrooms to capture video recordings of the lectures. This can help in two ways. One is that those video lectures can be made available to the students as well so that the absentee students can attend the lectures later as well. The students who attended the class, in fact, can also watch the recordings at a later time so that they can revise the lessons. But the second and the most important use of those recordings can be to assess the academic emotions or attentiveness of the students in class, as discussed in the earlier section. Applying various facial recognition tools, the concerned team can assess students' facial expressions and assess their boredom or attention. The same thing can be done by using various NLP (Natural Language Processing) and speech recognition tools to the transcripts of the video recordings (Cook et al., 2018). After an assessment of all hidden factors, various clustering techniques can be deployed to cluster the students in various groups such as 'brilliant performers', 'mediocre' and 'high risk' students, and necessary steps or actions can be taken by the administration to cater them with special care to improve performance.

8.5.2 Improvement of Students' Performance with the Use of AI in Online Platforms

In the case of online learning platforms, various data mining models have been developed that can tell us 'When' or 'what' actions are to be taken to improve the performance of a particular student. Since in online courses, the concerned team has access to each student personally, it becomes easier to identify which students are disengaging and when in a particular course. To meet this purpose, the various

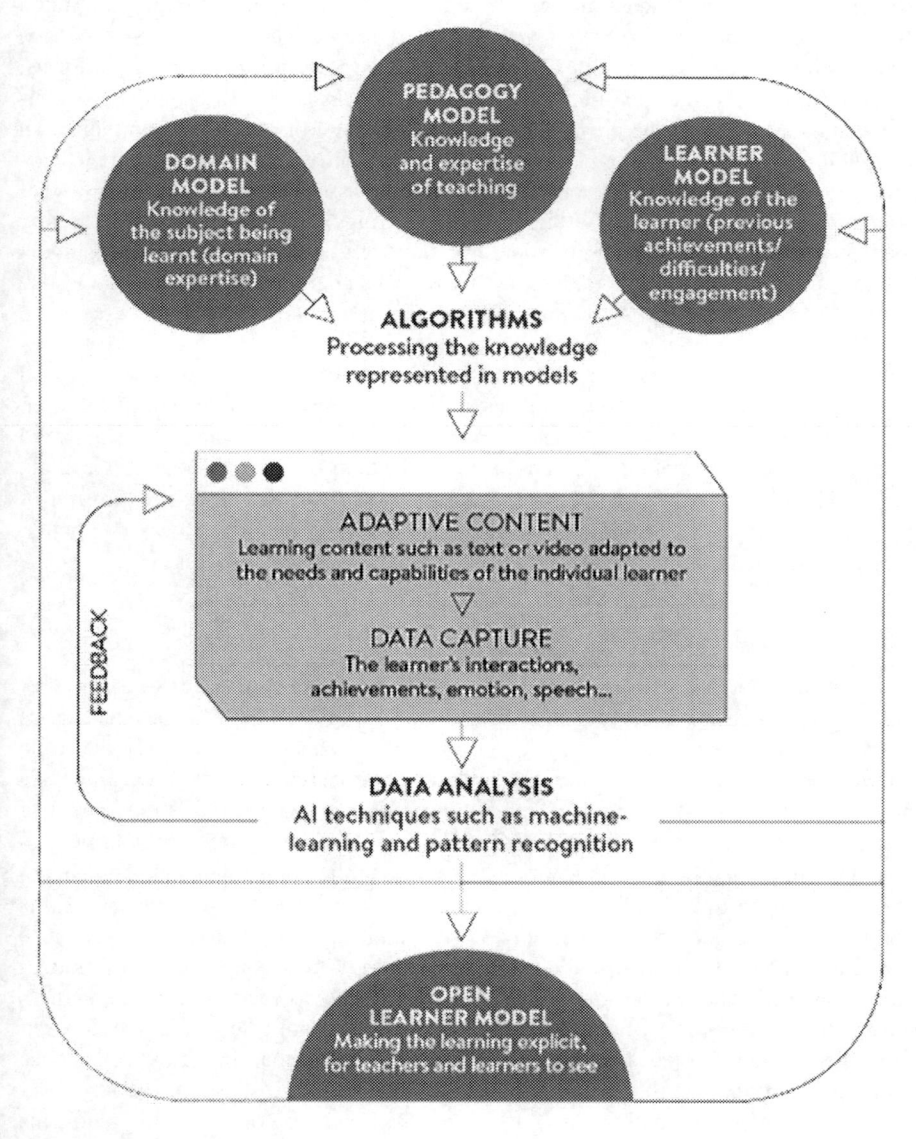

FIGURE 8.1 Flow Diagram of AI-Aided Online Learning Model.

effective detectors aforementioned can be deployed; the detectors give us something like a peek into a student's mind. It is possible to detect which subject or topic interests a student. It can also be determined what the student likes or dislikes, and moreover, when a particular student is losing interest in a topic taught, or when he or she is likely to quit a course (Whitehill et al., 2015). If the intention of a student to quit a course can be detected earlier, necessary interventions can be made by the concerned teaching faculty; the concerned team can improve the students' ability to be engaged with the course. For that, various inter-related lecture videos or texts can be suggested for a better understanding of the topic. The following flow diagram (cited from *www.raconteur.net/artificial-intelligence-is-the-next-giant-leap-in-education*) presents how AI can help students as well as instructors in online learning platforms.

Other than this, the idea of video recording and assessing the students' performance, as described in the previous section, for the case of offline environment, can be implemented in the online platforms too. In fact, online platforms with their various separate discussion forums provide more data to assess and have a clearer view of student's academic emotions. For instance, the writing behavior, raising of hands, and asking questions in the forum are source of data for assessing students' engagement in the course. Sometimes the interaction between a student and a course video is also analyzed to give a sense of student's engagement towards the course. For instance, the 'seeking and pausing' behavior of a student is directly related to his liking or disliking of the course materials. So, this can be used for analysis purposes as well (Atapattu and Falkner, 2018). Last but certainly not the least, the social media (Twitter, Facebook, etc.) participation of the students relating to the course can also be analyzed to discover the degree of engagement of the students in the course (Joksimović et al., 2015). The discussion clearly shows that online platforms provide vast exposure for the AI researchers and administrative bodies to assess students' performance better. And the administrative bodies can act accordingly as and when required.

8.6 CONCLUSION

This entire chapter discusses several uses of AI tools and techniques to assess students' performance and how it can improve with AI. In recent times, various studies have been conducted regarding *'blended learning'*, which means a mix of both offline and online learning environments. In the field of education, 'blended learning' is a new concept and needs to be developed and researched more towards the benefit of students' performance. We must mention that not only are students' academic performance improved by the applications of AI, but applications of AI help students to build their overall personality too. This has been described at the beginning of this chapter. For instance, various AI-aided co-curricular activity tools (Ng, 2021) have been developed, which help a student to focus with his hobbies outside of conventional studies as well. These tools encourage students to develop various activity-fondness apart from only studies, which in turn help the students not to become only a bookworm, but rather prosper as a better student and better human being as well. These tools can help students to develop noble characteristics like a sense of social responsibility, responsiveness and so on. For instance, various AI-aided quizzing platforms (e.g., QuizNext, Quillionz, etc.) enrich students with knowledge outside the book that is very much required for students' overall development. Various AI-integrated gaming apps like chess and puzzle-solving enrich the students' IQ and thinking capabilities. Although various research works regarding student performance analysis are going on and increasing day by day, the utilization of AI in the education sector for developing students' performance has not yet been implemented to an optimal extent. Thus, many academic institutions and online-course coordinators should apply these AI-aided inventions and tools to increasingly enhance students' performance.

REFERENCES

Aleven, V., Roll, I., McLaren, B. M., & Koedinger, K. R. (2016). Help helps, but only so much: Research on help seeking with intelligent tutoring systems. *International Journal of Artificial Intelligence in Education*, 26(1), 205–223. https://doi.org/10.1007/s40593-015-0089-1.

Artificial Intelligence in Education Market Size and Forecast. (2020, January). www.verified marketresearch.com.

Atapattu, T., & Falkner, K. (2018). Impact of lecturer's discourse for student video interactions: Video learning analytics case study of MOOCs. *Journal of Learning Analytics*, 5(3), 182–197. https://doi.org/10.18608/jla.2018.53.12.

Aulck, L., Nambi, D., & West, J. (2020). *Increasing enrollment by optimizing scholarship allocations using machine learning and genetic algorithms*. Proceedings of the 13th International Conference on Educational Data Mining.

Baradwaj, B., & Pal, S. (2011). Mining educational data to analyze students' performance. *International Journal of Advanced Computer Science and Applications (IJACSA)*, 2(6).

Bauer, A., Flatten, J., & Popović, Z. (2017). Analysis of problem-solving behavior in open-ended scientific-discovery game challenges. In X. Hu, T. Barnes, A. Hershkovitz, & L. Paquette (Eds.), *Proceedings of the 10th international conference on educational data mining* (pp. 32–39). Wuhan, China. Available at: https://pdfs.semanticscholar.org/fa02/2ca5d9b1f5364d5346ce0a6ee1cff0976840.pdf.

Biswas, G., Segedy, J. R., & Bunchongchit, K. (2016). From design to implementation to practice a learning by teaching system: Betty's brain. *International Journal of Artificial Intelligence in Education*, 26(1), 350–364. https://doi.org/10.1007/s40593-015-0057-9.

Botelho, A. F., Baker, R. S., & Heffernan, N. T. (2018). Improving sensor-free affect detection using deep learning. In E. Andre, R. Baker, X. Hu, M. Rodrigo, & B. duBoulay (Eds.), *Artificial Intelligence in education* (Vol. 10331, pp. 40–51). New York: Springer. https://doi.org/10.1007/978-3-319-61425-04

Bouchet, F., Harley, J., Trevors, G., & Azevedo, R. (2013). Clustering and profiling students according to their interactions with an intelligent tutoring system fostering self-regulated learning. *Journal of Educational Data Mining*, 5(1), 104–146. https://doi.org/10.5281/zenodo.3554613

Chatterjee, S., & Bhattacharya, K. (2020). Adoption of Artificial Intelligence in higher education: A quantitative analysis using structural equation modelling. *Education and Information Technologies*, 25, 3443–3463. doi: 10.1007/s10639-020-10159-7.

Chatterjee, S., Bhattacharjee, K., Tsai, C., & Agrawal, A. (2021). Impact of peer influence and government support for successful adoption of technology for vocational education: A quantitative study using PLS-SEM technique. *Journal of Quality and Quantity*. https://doi.org/10.1007/s11135-021-01100-2.

Chatterjee, S., Ghosh, S., & Chaudhuri, R. (2020). Knowledge management in improving business process: An interpretative framework for successful implementation of AI—CRM—KM system in organizations. *Business Process Management Journal*, 26(6), 1261–1281. doi: 10.1108/BPMJ-05-2019-0183.

Chatterjee, S., Majumdar, D., Misra, S., & Damaševičius, R. (2020). Adoption of mobile applications for teaching-learning process in rural girls' schools in India: An empirical study. *Education and Information Technologies*, 25, 4057–4076. doi: 10.1007/s10639-020-10168-6.

Chatterjee, S., Rana, N., & Dwivedi, Y. (2020). Social media as a tool of knowledge sharing in academia: An empirical study using valance, instrumentality and expectancy (VIE) approach. *Journal of Knowledge Management*, 24(10), 2531–2552. doi: 10.1108/JKM-04-2020-0252.

Chen, L., Chen, P., & Lin, Z. (2020). Artificial Intelligence in education: A review. *IEEE Access*, 8, 75264–75278. doi: 10.1109/ACCESS.2020.2988510.

Cook, C., Olney, A. M., Kelly, S., & D'Mello, S. (2018). An open vocabulary approach for estimating teacher use of authentic questions in classroom discourse. In *Proceedings of the 11th international conference on educational data mining* (pp. 116–126). Releigh, N.C: International Educational Data Mining Society. Available at: https://pdfs.semantic-scholar.org/1a75/8415f51474b9943faea5b506f98cadba1d15.pdf.

Davenport, T., & Kalakota, R. (2019). The potential of Artificial Intelligence in healthcare. *Future Healthcare Journal*, 6(2), 94–98. doi: 10.7861/futurehosp.6-2-94.

DeFalco, J. A., Rowe, J. P., Paquette, L., Georgoulas-Sherry, V., Brawner, K., Mott, B. W., . . . Lester, J. C. (2018). Detecting and addressing frustration in a serious game for military training. *International Journal of Artificial Intelligence in Education*, 28(2), 152–193. https://doi.org/10.1007/s40593-017-0152-1.

https://en.unesco.org/news/first-ever-consensus-artificial-intelligence-and-education-published-unesco. (n.d.).

Jiang, F., Jiang, Y., Zhi, H., Dong, Y., Li, H., Ma, S., . . . Wang, Y. (2017, 06). Artificial Intelligence in healthcare: Past, Present and Future. *BMJ*, 2, svn-2017. doi: 10.1136/svn-2017-000101.

Joksimović, S., Kovanović, V., Jovanović, J., Zouaq, A., Gasevic, D., & Hatala, M. (2015). What do cMOOC participants talk about in social media? A topic analysis of discourse in a cMOOC. In *Proceedings of the 5th international conference on learning analytics and knowledge* (pp. 156–165). Association for Computing Machinery. https://doi.org/10.1145/2723576.2723609.

Kabra, R., & Bichkar, R. (2011, December). Performance prediction of engineering students using decision trees. *International Journal of Computer Applications*, 36(11).

Lu, Y., & Hsiao, I. (2016). Seeking programming-related information from large scaled discussion forums, help or harm? In *Proceeding of the 9th international conference on educational data mining*. International Educational Data Mining Society. Available at: https://pdfs.semanticscholar.org/dc76/33aae5a8a99c2abac1c927617187cfefa7eb.pdf.

Moussavi, R., Gobert, J., & Sao Pedro, M. (2016). The effect of scaffolding on the immediate transfer of students' data interpretation skills within science topics. *Proceedings of the 12th international conference of the learning sciences* (pp. 1002–1005). Singapore: International Society of the Learning Sciences. https://doi.dx.org/10.22318/icls2016.157.

Ng, T. K. (2021). New interpretation of extracurricular activities via social networking sites: A case study of Artificial Intelligence learning ata secondary school in Hong Kong. *Journal of Education and Training Studies*, 9, 49–60. doi: 10.11114/jets.v9i1.5105.

Ogan, A., Walker, E., Baker, R., Rodrigo, M. M., Soriano, J. C., & Castro, M. J. (2015). Towards understanding how to assess help-seeking behavior across cultures. *International Journal of Artificial Intelligence in Education*, 25(2), 229–248. https://doi.org/10.1007/s40593-014-0034-8.

Park, J., Denaro, K., Rodriguez, F., Smyth, P., & Warschauer, M. (2017). Detecting changes in student behavior from clickstream data. In *Proceedings of the 7th international conference on learning analytics and knowledge* (pp. 21–30). Association for Computing Machinery. https://doi.org/10.1145/3027385.3027430.

Park, J., Yu, R., Rodriguez, F., Baker, R., Smyth, P., & Warschauer, M. (2018). Understanding student procrastination via mixture models. In *Proceedings of the 11th international conference on educational data mining*. International Educational Data Mining Society. Available at: https://pdfs.semanticscholar.org/67a4/027404642fad7f9baf0c5d-76dea14c99f563.pdf.

Pennebaker, J. W., Boyd, R. L., Jordan, K., & Blackburn, K. (2015). *The development and psychometric properties of LIWC2015*. Austin, TX: University of Texas at Austin. LIWC2015 Development Manual.

www.raconteur.net/artificial-intelligence-is-the-next-giant-leap-in-education. (n.d.).

Roll, I., & Winne, P. (2015). Understanding, evaluating, and supporting self-regulated learning using learning analytics. *Journal of Learning Analytics*, 2(1), 7–12. https://doi.org/10.18608/jla.2015.21.2.

Sao Pedro, M., Jiang, Y., Paquette, L., & Baker, R. S. (2014). Identifying transfer of inquiry skills across physical science simulations using educational data mining. In *Proceedings of the 11th international conference of the learning sciences* (pp. 222–229). Boulder, CO: International Society of the Learning Sciences. https://pdfs.semanticscholar.org/0ca7/cfc8e4575225162dd02c9ff64372158b030b.pdf.

Segedy, J. R., Kinnebrew, J. S., & Biswas, G. (2015). Using coherence analysis to characterize self-regulated learning behaviours in open-ended learning environments. *Journal of Learning Analytics*, 2(1), 13–48. https://doi.org/10.18608/jla.2015.21.3.

Shahiri, A., Husain, W., & Rashid, N. (2015). A review on predicting student's performance using data mining techniques. *Procedia Computer Science*, 72, 414–422. doi: 10.1016/j.procs.2015.12.157.

Subrahmanyam, V., & Swathi, K. (2018). *Artificial Intelligence and its implications in education*. International Conference on Improved Access to Distance Higher Education Focus on Underserved Commuinities and Uncovered Regions, Kakatiya University, Warangal, India.

Toth, K., Rolke, H., Greiff, S., & Wustenberg, S. (2014). Discovering students' complex problem-solving strategies in educational assessment. In *Proceedings of the 7th international conference on educational data mining*. London. Available at: https://pdfs.semanticscholar.org/1baf/8106de777bd369cbc7eb4e1d8c6689aed420.pdf.

Whitehill, J., Williams, J. J., Lopez, G., Coleman, C. A., & Reich, J. (2015). Beyond prediction: First steps toward automatic intervention in MOOC student stopout. In *Proceedings of the 8th international conference on educational data mining*. Madrid, Spain: International Educational Data Mining Society. https://dx.doi.org/10.2139/ssrn.2611750.

Yadav, S., Singh, S., Bora, A., & Thakur, S. (2016, November). Educational recommendation and tracking system. *International Journal of Scientific & Engineering Research*, 7(11), 1443–1447.

Zirawaga, V., Olusanya, A., & Maduku, T. (2017). Gaming in education: Using games as a support tool to teach history. *International Journal of Scientific & Engineering Research*, 8(15).

9 Artificial Intelligence-Based Tools in Research Writing
Current Trends and Future Potentials

Donnie Adams and Kee-Man Chuah

CONTENTS

9.1 INTRODUCTION

Research writing or academic writing has become a major component of learning and teaching across many disciplines in higher education (Strobl et al., 2019). The most common purpose of academic writing is to describe ideas or research findings and to convince the 'audience', which means the people reading the paper, such as instructors, professors and colleagues in the field, that the explanation is accurate. Research writing is strongly influenced by epistemological and communicative styles of the discipline (Poe et al., 2010) and genres used for teaching (Hyland, 2007; Nesi & Gardner, 2012), which result in a great range of procedures and skills involved in writing. These academic writing procedures and skills has become particularly challenging for both non-native English speakers and international students (Campbell, 2019). The use of English as the medium of instruction in many university graduate programmes has caused students difficulties with grammar, lexis and syntax (Singh, 2019). These difficulties worsen when students are faced with the

challenge of organizing ideas, sustaining arguments, defending claims and synthesizing ideas for their research writing (Belcher, 1994). Regrettably, very little is being
done to prepare and equip graduate students in academic writing.

Research writing is a complex, indispensable, and integrative task that has a major
influence on students' academic success (Rahimi & Zhang, 2018). Postgraduate students must fulfill various academic literacy demands such as individual project work,
group projects, discussion group works and research dissertations or theses as part of
their master's or PhD programme. In some cases, these students need to write and publish in citation indexed journals (CIJs) as part of their requirements to graduate (Adams
et al., 2021). The greatest challenge among these tasks is producing written works free
of language errors. These errors often create a negative first impression (Singh, 2019).
Brown (2008), suggesting that students do not only face challenges in terms of vocabulary and grammar, but that they also have an inadequate understanding of academic
writing standards and expectations from their lecturers. Approaches to improve writing proficiency have been widely discussed and assessed in the past (Graham & Perin,
2007; Strobl et al., 2019). Among the most effective measures for novice writers are
extended practice, individualized feedback and strategy instruction (Allen et al., 2016).
These measures, however, require considerable amounts of time, efforts and costs.

Questions begin to arise around to what extent research writing and training
can be supported electronically (Crossley & McNamara, 2016). The emergence of
Artificial Intelligence (AI) technologies have triggered a tremendous interest among
educational technologists. One area that has received much attention is AI-based
tools, developed to assist researchers in the writing process. AI-powered writing tools
aim to not only ease the process of research writing but also to enhance the quality
of critical analysis particularly in the aspect of literature review and language style.
Despite the trends in adopting AI components within research writing, there are
still-limited efforts to examine its implementation, strengths and weaknesses. This
chapter reviews some of the key studies concerning AI in research writing as well as
an extensive review of existing tools by covering their AI-based features, affordances
and constraints. The chapter also includes a discussion on the future potentials of AI
implementation with regard to research writing. This chapter would serve as a good
reference to uncover the hidden potentials of AI-based tools in assisting students to
produce high-quality research writing.

9.2 CURRENT TRENDS OF AI COMPONENTS IN RESEARCH WRITING

This section looks at trends of AI components by sketching their effect on research
writing. The term 'tool' in this chapter refers to writing technologies ranging from
programs, applications and platforms (Cotos, 2015). Artificial Intelligence (AI) on
the other hand refers to an automated device with human intelligence processes
such as learning, interpreting and self-correction (Popenici & Kerr, 2017). AI is able
to analyze and do tasks as a human does (Strobl et al., 2019). As monitoring students' writing and providing timely, valuable and productive feedback is becoming
too time-consuming and requires high levels of labor work, writing technologies
are being developed to facilitate these needs (Lim & Phua, 2019). Computer-based

applications are increasingly being used as alternatives to facilitate research writing such as automated essay scoring (AES), automated written corrective feedback (AWCF) and automated writing evaluation (AWE) (Allen et al., 2016). These AI-powered writing tools can assist students to learn and develop their research writing skills.

New writing applications offer time-saving additions and flexibility to the writing curriculum (Koltovskaia, 2020). For example, AES has proven to reduce educators' grading workload. AWE is capable of providing both formative and summative feedback to students (Foltz et al., 2013). AWCF is an effective tool to improve student research writing engagement (Nazari et al., 2021). Thus, AI-powered writing tools now play an increasingly valuable role in research writing (Zhang, 2020). Apart from providing information about learning, AI applications can provide a thorough instructional practice and assist students in their research writing progress (Zawacki-Richter et al., 2019). Perhaps the most salient contribution of AI-powered writing tools are seen in assessment, feedback, tutoring and content generation for teachers and students (Nazari et al., 2021).

The literature shows that the AI-powered writing tools that have been used to automatically analyze research writing can be largely classified into four components, namely rule-based, corpus-based detection, natural language processing (NLP) and deep learning, or neural network. The rule-based tool is a grammar checking system commonly used to suggest and check common sentence structures and academic words in research articles. It provides detailed explanation of flagged errors thus making it an extremely helpful computer aided language learning approach (Soni & Thakur, 2018). Corpus-based tools use text summarization to identify important sentences from a substantial amount of information and produce a concise and fluent summary while preserving key information and overall meaning (Extance, 2018). NLP is the ability of a computer programme to detect the sentiment or tone in writing. The programme takes real-world input, processes it and attempts to understand human language (referred to as natural language) as it is spoken and written (Panesar, 2020). Deep learning, or neural network, programmes are capable of reading scientific papers and rendering a summary in a sentence or two. It is useful for helping writers get a preliminary sense of what these scientific papers are about and enables better decision-making in their research writing (Dangovski et al., 2019). Table 9.1 indicates the key areas where AI components are incorporated in the scope of research writing.

The remaining sections of this chapter will present an extensive review of AI in research writing as well as its features, affordances and constraints.

9.3   REVIEW OF STUDIES ON AI IN RESEARCH WRITING

Studies pertaining to the adoption of Artificial Intelligence in research writing are centred on the goal of not only increasing the efficiency in writing research materials but also providing useful information for researchers to make informed decisions. AI-based tools for specific purposes are rooted in the development of intelligent tutoring systems since the late 1980s (Richardson, 1988). The foundation of intelligent tutoring systems is the use of multiple independent modules to carry out a given

TABLE 9.1

The Key Areas Where AI Components are Incorporated in the Scope of Research Writing.

AI Approaches	Scope	Usages
Rule-based	Identification or detection based on specific linguistics rules (grammar)	• Grammar checker • Spelling error detection • Suggested word lists
Corpus-based	Summarization of corpus in creating patterns of usage and identifying collocations.	• Texts summarizer • Word concodence • Academic phrasebank
Natural Language Processing	The processing of human language for more complex linguistic analysis.	• Sentiment analysis • Tone/Diction analysis
Deep learning	Deeper analysis through neural networks or advanced AI with unsupervised training.	• Textual analytics • Advanced summarizer • Real-time intelligent tutors

function. For example, an expert module is used to store domain knowledge and captures underlying intelligence behaviour, while cognitive expert modules are created to simulate the actual human problem-solving capability within a domain or subject matter (Anderson, 1988, 1993). Essentially, AI-based tools are inspired by the idea of enhancing the learning of good writing through various affordances that are included within each tool, without replacing the role of the writer in the process (Crompton & Song, 2021; Khan et al., 2020; Lu, 2019).

In recent years, AI-based tools for research writing are geared towards facilitating the process of conducting common research tasks (such as literature review and data visualization) as well as providing continuous feedback to authors in improving their writing and proof checking (such as similarity index and spelling checkers). In a randomized controlled trial on 120 postgraduate students, Nazari et al. (2021) found that AI-based writing tools were able to improve the students' learning behaviour and attitudinal technology acceptance. The group with a specific focus on the tools were shown to produce better writing, as they are constantly given formative feedback. While the tools may at times offer inaccurate suggestions, they were largely useful in motivating the learners to be more confident in their research writing. Cotos et al. (2020) focused on the similar scope by developing the Research Writing Tool (RWT) that is capable of examining research writing through different forms of automated feedback. Grounded by the corpus-based approach, each written work is checked, and authors would be given both macro-level (structure, cohesion, etc.) and micro-level comments (sentence structures, word choice, etc.) for them to consider. The suggestions given by RWT offer great help for novice writers who may not be aware of a specific style of research writing within a discipline.

In relation to the feedback provided by AI-powered tools, some studies have shown how beginning writers are favouring automated feedback as compared to feedback provided by humans. Wang (2020) found such pattern in his study on 188 university students in China taking an advanced English reading and writing course, which was predominantly about writing scientific papers. The experiment revealed that the students preferred the feedback given through automated essay evaluation systems. The perceived effectiveness of the system was also higher than the teacher ratings. The three selected systems (Pigai, iWrite, and Awrite) were based on natural language processing in which a large corpus was used for the system to learn and identify the patterns of good writing. Through the systems' algorithmic matching and cross-examination, detailed individualized feedback is given to the students along with holistic scoring. Interestingly, Wang (2020) also discovered that because of the individualized feedback, students' independent learning ability and writing ability were vastly improved. This finding shows that AI-based tools are not merely to save time or make the process more convenient, but at the same time they are helping novice writers to learn the proper way of writing.

Moreover, the use of AI in research writing tools helps deal with some humanistic factors in writing, such as motivation and anxiety. It is widely known that most students or novice researchers lack the motivation to write scientifically due to the prevalent perception of the difficulty in producing research writing (Tansomboon et al., 2017). The AI-based tools are capable of offering unbiased feedback and responses for students to revise their written work. Tansomboon et al. (2017) explained how NLP tools can be capitalized to analyze student writing and select the most optimal guidance for each area of concern. Strobl et al. (2019) also highlighted how Automated Writing Evaluation (AWE) and Automated Essay Scoring (AES) systems are improving students' writing proficiency while increasing opportunities for formative feedback, rather than summative. Hence, students who get real-time support in the writing process (especially in scientific writing) are more likely to stay motivated and reduce their anxiety in writing.

The development of AI in research writing is not restricted to end-user tools alone. There are also studies that examine different ways to improve the AI processes from the rule-based engine to the more intelligent natural language processing. One example is the work by Sirbu et al. (2018) that specifically applied the ReaderBench framework to assess over 15,000 essays in order to generate 1,200 textual complexity indices. These indices were then grouped into four categories: word complexity, local cohesion, global cohesion and normalized word counts related to sentiment polarity (positive and negative). By creating these indices, an extensible rule-based engine could be developed in order to provide personalized feedback to users. Shum et al. (2016) also initiated the same efforts in going beyond the typical information retrieval metrics by introducing NLP-based writing analytics. The analytical tool harnesses the power of NLP approaches to examine various aspects of texts, from sentence difficulty to the cohesion of a writing. Additionally, Ullman (2019) examined the use of both machine learning and rule-based approaches in analyzing reflective writing. The machine learning model used in his study managed to identify five out of eight categories of reflective writing automatically. Although the accuracy was slightly lower than manual analysis, his work has provided the initial foundation on reflective writing analytics. These studies indicate that AI in writing is not confined only to software development; there is also an increasing interest in expanding the capability of AI approaches.

However, it is noteworthy that studies on adaptive systems for teaching and learning in higher education do not seem to focus on research writing thus far. The use of the adaptive system is still related to providing academic advice to students (Alfarsi et al., 2017), supporting university career services (Nguyen et al., 2018) and enhancing learning management systems with features that can recommend personalized contents to students (Kose & Arslan, 2016). Adaptive systems for research writing can be said to be still in their infancy but with great potential, as they allow students to obtain intelligent feedback rather than rule-based outputs. Ocharo and Hasegawa (2018) introduced an adaptive system for academic writing, which was developed on the basis of cognitive apprenticeship theory. It contains modules for modelling, coaching and scaffolding. From the students' initial drafts to the final drafts, the system is capable of offering feedback and guidance adaptive to the cognitive needs of the students. The adaptive interface is shown to students based on the initial skill-level estimation engine, whereby students are classified into novice, intermediate and advanced. Necessary guidance is then presented to students based on the estimated skill level.

This review of the studies related to the use of AI in research writing has demonstrated not only its wide-ranging adoption but also its importance in guiding students and novice writers. Most of the tools reviewed were not commercially released but have provided valuable input on various uses of AI approaches in guiding research writing. However, due to the advancement of cloud computing and web development, AI-based tools for research and academic writing are now easily accessible. The following section will present some of the popular AI-based tools in terms of their features and affordances.

9.4 FEATURES AND AFFORDANCES OF AI-BASED TOOLS IN RESEARCH WRITING

The AI-based tools for research writing released publicly either in the form of freeware or premium software can be largely categorized into three scopes: language and mechanics, text summarizing and synthesizing, and formatting and typesetting.

9.4.1 Language and Mechanics

NLP and rule-based engines allow tools for language checking to be developed. These tools help users identify errors in language (grammar and sentence structures) and mechanics (such as punctuation, capitalization, abbreviations). The first popular tool in this category is Grammarly. It comes in free and premium versions and is largely used as a language checker. It can be used as a plugin on a browser or popular word processing software such as Microsoft Word (Fitria, 2021). As shown in Figure 9.1, Grammarly has several metrics that indicate the writing quality from correctness (accuracy in terms of spelling and grammar) to engagement and clarity. Driven by corpus-based databases and NLP, Grammarly's capability has shown tremendous improvement thanks to the increasing corpus size for deep learning to be implemented.

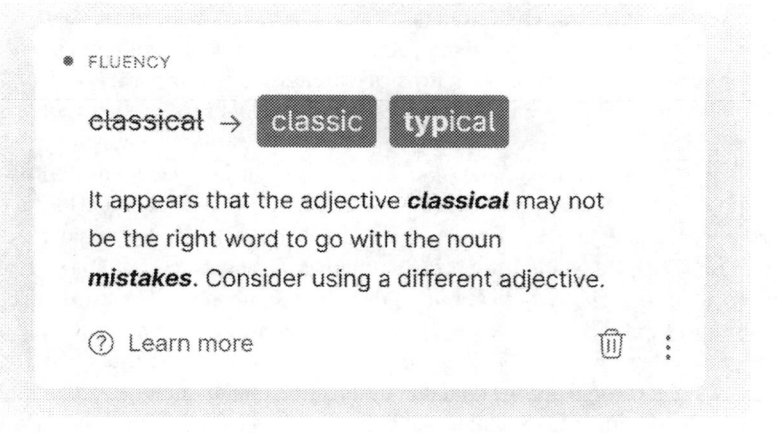

Mispellings and grammatical errors can effect your credibility. The same goes for misused commas, and other types of punctuation . Not only will Grammarly underline these issues in red, it will also showed you how to correctly write the sentence.

Underlines that are blue indicate that Grammarly has spotted a sentence that is unnecessarily wordy. You'll find suggestions that can possibly help you revise a wordy sentence in an effortless manner.

But wait...there's more?

Grammarly Premium can give you very helpful feedback on your writing. Passive voice can be fixed by Grammarly, and it can handle classical word-choice mistakes. It can also help with inconsistencies such as switching between e-mail and email or the U.S.A. and the USA.

It can even help when you wanna refine ur slang or formality level. That's especially useful when writing for a broad audience ranging from businessmen to friends and family, don't you think? It'll inspect your vocabulary carefully and suggest the best word to make sure you don't have to analyze your writing too much.

- commas, · Remove the comma
- punctuation . · Remove a space
- , · Add the word(s)
- showed · Change the verb form
- to correctly write the sentence · Ungist the infinitive
- a sentence that is unnecessarily w... · Remove wordiness
- possibly · Remove redundancy
- revise a wordy sentence in an eff... · Change the wording
- ... · Remove the ellipsis
- very helpful · Choose a different word
- classical · Change the word

FIGURE 9.1   The Grammarly Document Checking Interface.

FIGURE 9.2   The User-Friendly Feedback Boxes in Grammarly.

Grammarly's main affordance is in its simplicity in offering necessary language checking (Syafi'i, 2020). Another affordance is in terms of its accuracy. It has been studied to be more accurate than other language tools (which are built using simple rule-based engines). Sahu et al. (2020), for example, compared Grammarly with four other apps, and noted Grammarly to be the most accurate one, although it still has problems semantics and complex sentence structure. Grammarly also provides explanations for suggestions that it gives to the user, as shown in Figure 9.2. This feature helps users decide whether changes are necessary so they do not blindly accept all the suggested corrections by Grammarly.

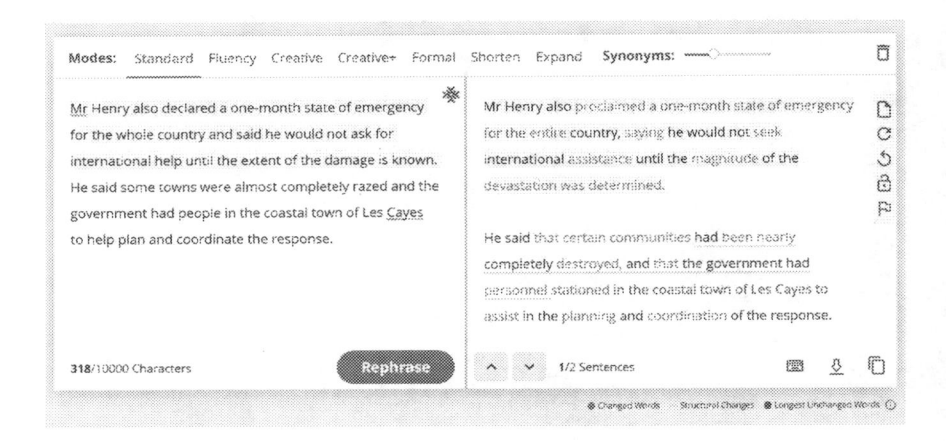

FIGURE 9.3   The Web Version of QuillBot.

Another popular AI-tool for language is Quillbot, which is known as an "intelligent" paraphrasing tool. Similar to Grammarly, it is offered in two versions: free and premium. The free version has a limit in terms of the number of words that can be input. It was first introduced as a paraphrasing helper but it has now incorporated a summarizer and grammar checker as well. Primarily, its strength is still in terms of paraphrasing, as the tool comes with several affordances that does not exist in other tools, as shown in Figure 9.3. The first affordance is to select the modes of paraphrasing; whether the users want to focus on fluency or more creative use of language. Quillbot is capable of paraphrasing sentences provided into different styles. The second affordance is the feature to decide the number of words that users want to change (synonym percentage). This option is helpful as it allows users to control the suggested outputs so that not all words are changed randomly to maintain the original meaning. The third affordance is the feature to expand and shorten what was provided in the original text. A text can be expanded and shortened without changing the meaning, allowing users to find different possibilities of writing the same concept. Finally, the fourth interesting affordance is its ability to provide other analytical tools, namely, comparing paraphrased texts according to different modes, and word and sentence count.

While QuillBot seems to offer many useful features, it is worth noting that most of the affordances mentioned earlier are only available in the premium version. Studies that examine QuillBot's effectiveness and usefulness are also limited. In a study by Inayah and Sulistyaningrum (2021), they found QuillBot to be helpful in helping students overcome difficulties in paraphrasing, as it is often perceived as a challenging task in research writing. Nevertheless, they also noted that due to the convenience of generating paraphrased texts, students tend to directly copy the output without confirming whether the meaning is retained.

The two AI-based tools for language and mechanics mentioned are still very useful especially for researchers or writers who may be pressured to produce a writing in a short period of time. These tools would be able to assist them to be more efficient though they should be vigilant in cross-checking the accuracy of the tools.

9.4.2 TEXT SUMMARIZER AND SYNTHESIZER

One of the most challenging tasks in research writing is in performing a literature review. In order to produce good research writing, a solid review of previous studies and related literature is necessary. In this case, one popular tool is Scholarcy. The AI-powered article summarizer can quickly scan an article and provide the needed synopsis beyond the abstracts (refer to Figure 9.4). It reads the articles and breaks them down into easily readable sections so that users can assess how important the article is without having to read the whole text (Klucevsek & Brungard, 2020). Although it will not automatically write the review for the users, this tool is widely used for filtering articles and helping researchers to efficiently pick the most relevant articles for their research writing.

The effectiveness of this tool remains debatable since it still largely depends on the keywords used in the articles. For example, if a section in an article does not follow the usual convention of research writing, Scholarcy would have problems summarizing the correct information. An affordance of this tool is that it also comes with a reference and literature review manager. Users can manage downloaded or bookmarked articles easily in the app and get the synopsis of each article directly as shown in Figure 9.5. Scholarcy does seem to offer intriguing affordances to researchers and students, particularly in its ability to rapidly scan through texts and offer necessary output via its deep reference mining feature.

Iris.ai is another AI-based tool that is capable of summarizing and synthesizing texts. It boasts itself as the world-leading AI engine for scientific text understanding, which can be applied for the purpose of literature reviews, data extraction and tasks that require the processing of a large number of documents. Offered in free and premium packages, Iris.ai is still pretty new in the market but has built its reputation in its corpus-based NLP engine and intuitive visualizations, as shown in Figure 9.6 and Figure 9.7.

The primary affordance of Iris.ai is in its capability to transform what was found in research articles into neatly presented categorization and mapping (Extance,

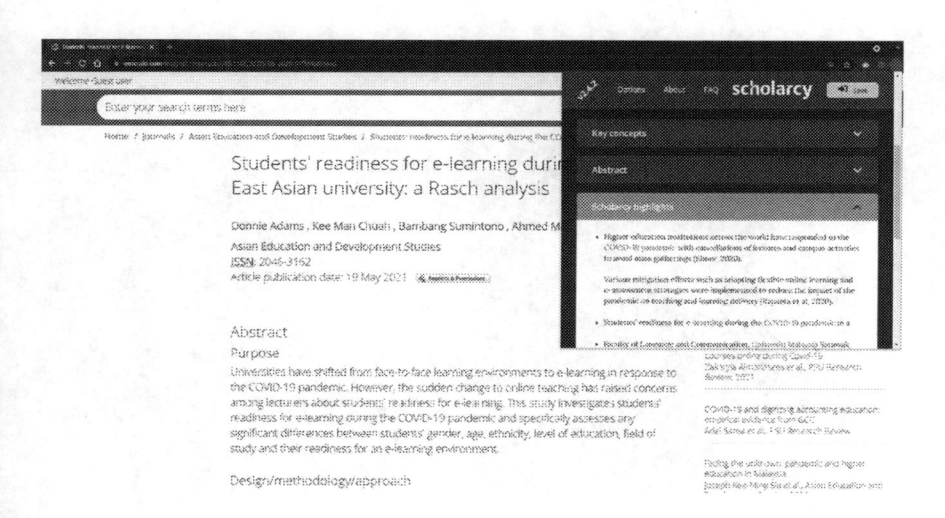

FIGURE 9.4 The Scholarcy Plug-In on Chrome Browser.

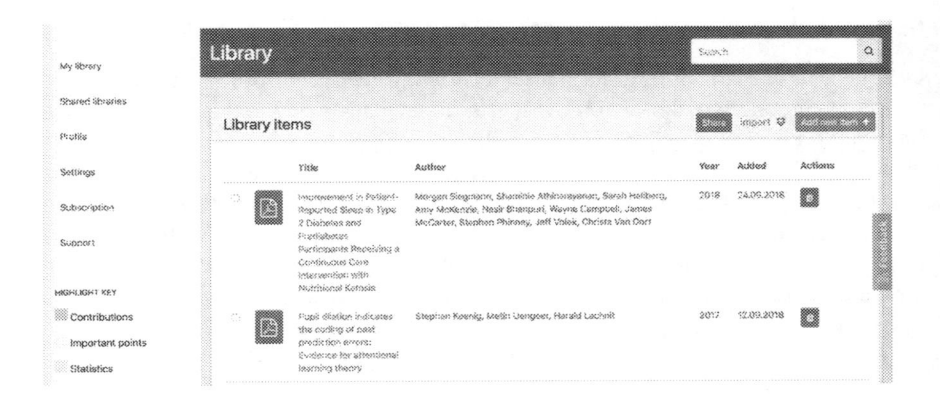

FIGURE 9.5   The Literature Review Manager in Scholarcy.

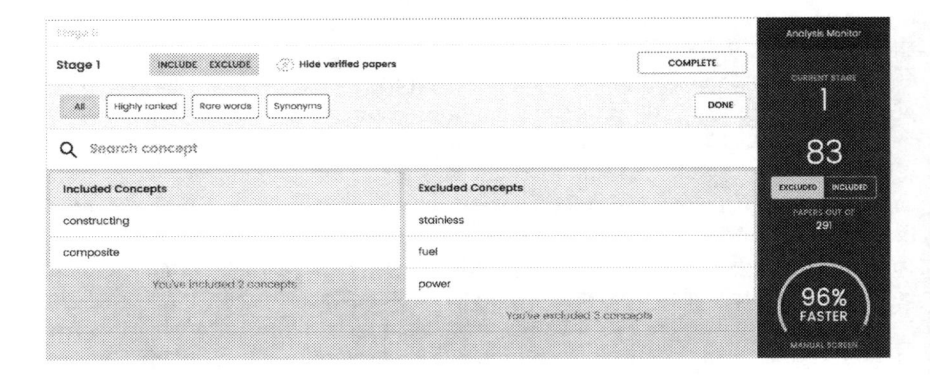

FIGURE 9.6   The Iris.ai Literature Search and Summary.

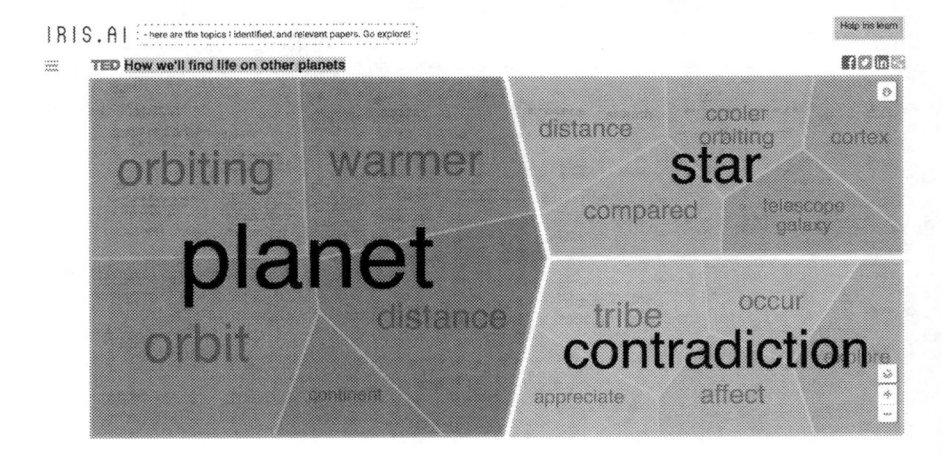

FIGURE 9.7   The Visualization of Identified Research Papers Based on Concepts.

2018). It also allows users to not only narrow down the reading list but also see each paper's relevance score to a stated problem statement. This feature is indeed powerful in assisting researchers to save more time in finding the relationships between documents. Another affordance of this tool (the premium version) is its pre-included data with access to more than 200 million papers. In other words, researchers do not have to spend time finding the papers or data from different databases as the search can be done directly through Iris.ai. All the filtered research papers are then compiled within Iris.ai for deeper checking, especially in terms of the parameters set by the users. Due to its pricing, Iris.ai seems to be more affordable to researchers rather than students. Although it provides a discounted rate for students, the price range is still relatively high. Thus, the opportunity for students to fully utilize this tool is limited. The free version is still sufficient for students to perform a basic literature search and obtain necessary summaries.

9.4.3 Typesetting

Another main problem in research writing is in terms of fulfilling different format and typesetting pre-determined by various publishers. While LaTeX tool is widely used in academia for handling typesetting (Baramidze, 2013), its programming-like commands and interface tend to be confusing to beginners. Typeset.io is created to overcome this problem. This web-based tool allows users to select a format of their choice (for example, paper format for a specific journal in major databases like IEEE or Springer) and upload their content to be converted to the required format instantly. As shown in Figure 9.8, once the user selects a paper format for a journal, they can immediately upload their content in Microsoft Word to be converted into the right format. It also allows editing to be made online directly.

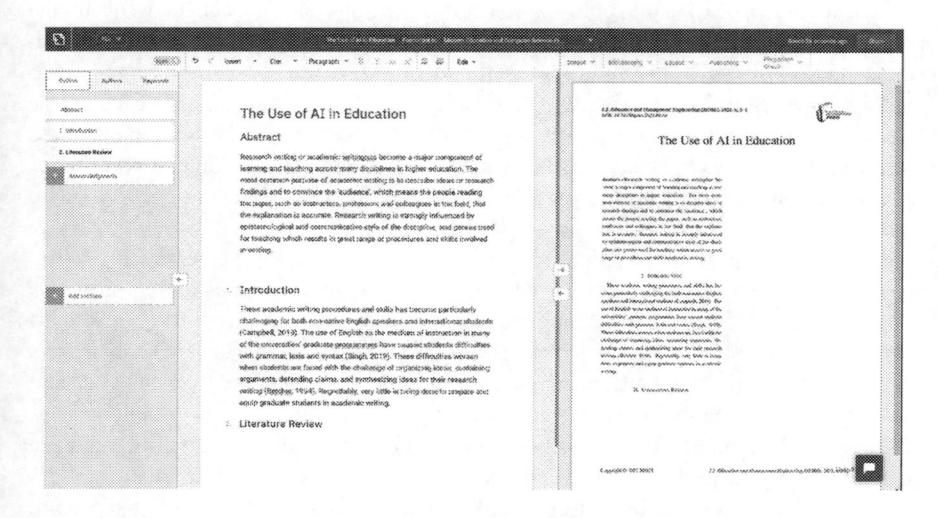

FIGURE 9.8 Typeset.io Sample Template for a Journal.

Typeset.io is by far the most convenient tool for typesetting, and though it does not use high-end Artificial Intelligence, its ability to detect key sections and learn to match them within the uploaded content can still be considered useful in research writing. Its large selection of different journals' format requirements is indeed another affordance that can benefit researchers and students.

9.5 FUTURE POTENTIALS

The tools reviewed in the previous sections have provided valuable insights into how AI-based tools can assist research writing. The process of writing any research publication is not simple and straightforward for many beginners and students. Tools that make use of AI and its corresponding technologies and approaches are beneficial. Future development and implementation of AI in research writing should focus on several potential areas. The first area is developing tools grounded on pedagogical features in which guidance is given in a more instructor-like manner (Shum et al., 2016). As learners require advice to produce good writing, suggestions and comments can be offered through a method that considers pedagogy. For instance, instead of directly suggesting a sentence to be changed, the tool can provide some guiding questions that would inform the system to offer adaptive responses to the learners.

Additionally, future development of AI-based tools for research writing should prioritize macro-level issues and process-oriented writing (Strobl et al., 2019). Most of the tools thus far focus on micro-level issues such as spelling errors, sentence structure and mechanics. There is a huge potential to optimize the use of AI in guiding learners to manage issues concerning cohesion and coherence, argumentative quality and diction or tone. As research writing must be unbiased and objective, the aspect of sentiment analysis in these tools should be given more attention as well. This would help not only students but also researchers to be aware of the claims or statements made.

Developers of AI-based tools for research writing can also consider the incorporation of conversational bots or chatbots. Although it may seem like a mismatch since chatbots are speaking and reading, the informal nature of chatbots can enhance writing instructions by lowering the affective filters (Lin & Chang, 2020). Since writing is often perceived as a difficult skill, using chatbots would be able to simulate the presence of a teacher in assisting them in writing. This inclusion can potentially reduce their anxiety and lead to positive emotional impacts in learning how to write research papers.

The future potentials of enhancing AI adoption in developing research writing tools is driven by the development in AI itself. As AI technologies and approaches become more advanced and cheaper, the developed tools can, therefore, deal with more complex problems in research writing. There are, however, several constraints that need to be given attention. Firstly, no matter how sophisticated an AI-based tool would be, the elements of semantics and pragmatics are still dependent on contexts, which can change drastically. Machine learning would need to cope with the changes in how certain words or sentences are used to convey certain meanings. As emphasized by Labov (2018), many words with newly found meanings or usages are no longer restricted by geographical constraints. The usage of "avant garde" terms makes it even more difficult for machine learning to "decode" the correct meaning. This shortcoming highlights the need for more in-depth research using more robust algorithms in dealing with the complexity of natural language use. Deep learning beyond the Bayesian methods such as deep belief network (Huang et al., 2019) is showing great promise in understanding language use from large corpus. Secondly, AI-based research writing tools need to cope with requirements of different fields, as each field may have a set of expectations to be addressed. For example, research writing in the medical field is quite different from that in humanities. Thirdly, it is pivotal for developers of AI-based research writing tools to offer formative feedback, which are meaningful, rather than a repetitive pre-determined feedback. Meaningful feedback is indeed necessary for the development and enhancement of research writing skills. Despite the constraints, it would be interesting to see how AI continues to develop in the future and play its role in the area of research writing.

9.6 CONCLUSION

The area of research writing or academic writing in general may seem trivial but its impact in promoting scholarly sharing and dissemination of knowledge cannot be taken lightly. This chapter has provided useful insights on how knowledge and technicalities of AI can lend a hand in elevating the effectiveness of research writing tools. More empirical research should be conducted to validate the tools as well. Additionally, future potentials of AI implementation for complex research writing are also presented. Despite the excitement of easing the process of research writing, this area should also give more weight to ethical issues such as plagiarism and research fabrication. The development of better cognitive processes for AI intelligent systems would definitely help to prevent academic dishonesty or unethical practices in research writing. It is important to counter the notion that AI-based tools are going to turn research writing into an automatic factory-like process without much intellectual consideration. In terms of pedagogical implications, the AI-based tools should not be used as a direct replacement of instructors' role in offering personalized and meaningful feedback to the learners. They are to be used cautiously as supplementary tools to assist instructors in managing writing instruction and effective assessment of research writing. In essence, learners still must master the knowledge and skills of research writing and use available tools to facilitate the process.

REFERENCES

Adams, D., Thien, L. M., Chuin, E. C. Y., & Semaadderi, P. (2021). The elusive Malayan tiger 'captured': A systematic review of research on educational leadership and management in Malaysia. *Educational Management Administration & Leadership*, 1741143221998697.

Allen, L. K., Jacovina, M. E., & McNamara, D. S. (2016). Computer-based writing instruction. In C. A. MacArthur, S. Graham, & J. Fitzgerald (Eds.), *Handbook of writing research* (pp. 316–329, 2nd ed.). New York, NY: Guilford.

Alfarsi, G. M. S., Omar, K. A. M., & Alsinani, M. J. (2017). A rule-based system for advising undergraduate students. *Journal of Theoretical and Applied Information Technology*, 95(11), 2453–2465.

Anderson, J. R. (1988). The expert module. In M. Polson & J. Richardson (Eds.), *Foundations of intelligent tutoring systems* (pp. 21–53). Hillsdale, NJ: Lawrence Erlbaum Associates.

Anderson, J. R. (1993). *Rules of the mind*. Hillsdale, NJ: Lawrence Erlbaum Associates.

Baramidze, V. (2013). LaTeX for technical writing. *Journal of Technical Science and Technologies*, 45–48.

Belcher, D. (1994). The apprenticeship approach to advanced academic literacy: Graduate students and their mentors. *English for Specific Purposes*, *13*, 23–34.

Brown, L. (2008). The incidence of study-related stress in international students in the initial stage of the international sojourn. *Journal of Studies in International Education*, *12*(1), 5–28.

Campbell, M. (2019). Teaching academic writing in higher education. *Education Quarterly Reviews*, *2*(3), 608–614.

Cotos, E. (2015). Automated writing analysis for writing pedagogy: From healthy tension to tangible prospects. *Writing & Pedagogy*, *7*(2–3), 197–231.

Cotos, E., Huffman, S., & Link, S. (2020). Understanding graduate writers' interaction with and impact of the research writing tutor during revision. *Journal of Writing Research*, *12*(1), 187–232. https://doi.org/10.17239/jowr-2020.12.01.07.

Crompton, H., & Song, D. (2021). The potential of Artificial Intelligence in higher education. *Revista Virtual Universidad Católica Del Norte*, *62*, 1–4. https://doi.org/10.35575/rvucn.n62a1.

Crossley, S. A., & McNamara, D. S. (2016). *Adaptive educational technologies for literacy instruction*. London: Routledge.

Dangovski, R., Jing, L., Nakov, P., Tatalović, M., & Soljačić, M. (2019). Rotational unit of memory: A novel representation unit for rnns with scalable applications. *Transactions of the Association for Computational Linguistics*, *7*, 121–138.

Extance, A. (2018). How AI technology can tame the scientific literature. *Nature*, *561*(7722), 273–275.

Fitria, T. N. (2021). Grammarly as AI-powered English writing assistant: Students' alternative for writing English. *Metathesis: Journal of English Language, Literature, and Teaching*, *5*(1), 65–78.

Foltz, P. W., Streeter, L. A., Lochbaum, K. E., & Landauer, T. K. (2013). *Implementation and applications of the intelligent essay assessor*. London: Routledge.

Graham, S., & Perin, D. (2007). A meta-analysis of writing instruction for adolescent students. *Journal of Educational Psychology*, *99*(3), 445–476.

Huang, Q., Yang, Y., & Cheng, M. (2019). Deep learning the semantics of change sequences for query expansion. *Software: Practice and Experience*, *49*(11), 1600–1617.

Hyland, K. (2007). Genre pedagogy: Language, literacy and L2 writing instruction. *Journal of Second Language Writing*, *16*(3), 148–164.

Inayah, N. A. M., & Sulistyaningrum, S. D. (2021). Employing online paraphrasing tools to overcome students' difficulties in paraphrasing. *Stairs English Language Education Journal*, *2*(1), 52–59.

Khan, A., Ayega, D., Aguilar, I., & Baker, R. (2020, November). The digital tutor and student engagement techniques: An intelligent way to engage students in ITS. In *Innovate learning summit 2020* (pp. 15–28). Association for the Advancement of Computing in Education (AACE). Available at: https://www.learntechlib.org/primary/p/218778/.

Klucevsek, K. M., & Brungard, A. B. (2020). Digital resources for students: Navigating scholarship in a changing terrain. *Libraries and the Academy*, *20*(4), 597–619.

Koltovskaia, S. (2020). Student engagement with automated written corrective feedback (AWCF) provided by Grammarly: A multiple case study. *Assessing Writing, 44,* 100450.

Kose, U., & Arslan, A. (2016). Intelligent e-Learning system for improving students' academic achievements in computer programming courses. *International Journal of Engineering Education*, *32*(1), 185–198.

Labov, W. (2018). The role of the Avant Garde in linguistic diffusion. *Language Variation and Change, 30*(1), 1–21.

Lim, F. V., & Phua, J. (2019). Teaching writing with language feedback technology. *Computers and Composition, 54*, 102518. https://doi.org/10.1016/j.compcom.2019.102518.

Lin, M. P. C., & Chang, D. (2020). Enhancing post-secondary writers' writing skills with a chatbot. *Journal of Educational Technology & Society, 23*(1), 78–92.

Lu, X. (2019). An empirical study on the Artificial Intelligence writing evaluation system in China CET. *Big Data, 7*(2), 121–129. http://doi.org/10.1089/big.2018.0151.

Nazari, N., Shabbir, M. S., & Setiawan, R. (2021). Application of Artificial Intelligence powered digital writing assistant in higher education: Randomized controlled trial. *Heliyon, 7*(5). https://doi.org/10.1016/j.heliyon.2021.e07014.

Nesi, H., & Gardner, S. (2012). *Genres across the disciplines. Student writing in higher education.* Cambridge: Cambridge Applied Linguistics.

Nguyen, J., Sánchez-Hernández, G., Armisen, A., Agell, N., Rovira, X., & Angulo, C. (2018). A linguistic multi-criteria decision-aiding system to support university career services. *Applied Soft Computing Journal, 67*, 933–940. https://doi.org/10.1016/j. asoc.2017.06.052.

Ocharo, H. N., & Hasegawa, S. (2018, July). Adaptive interface that provides modeling, coaching and fading to improve revision skill in academic writing. In *International conference on human interface and the management of information* (pp. 300–312). Cham: Springer.

Panesar, K. (2020). Natural language processing (NLP) in Artificial Intelligence (AI): A functional linguistic perspective. In S. S. Gouveia (Ed.), *The age of Artificial Intelligence: An exploration.* Delaware: Vernon Press.

Poe, M., Lerner, N., & Craig, J. (2010). *Learning to communicate in science and engineering: Case studies from MIT.* Cambridge, MA: MIT Press.

Popenici, S. A. D., & Kerr, S. (2017). Exploring the impact of Artificial Intelligence on teaching and learning in higher education. *Research and Practice in Technology Enhanced Learning, 12*(1).

Rahimi, M., & Zhang, L. J. (2018). Writing task complexity, students' motivational beliefs, anxiety and their writing production in English as a second language. *Reading and Writing, 32*(3), 761–786.

Richardson, J. J. (1988). *Intelligent tutoring systems.* Hillsdale, NJ: Psychology Press.

Sahu, S., Vishwakarma, Y. K., Kori, J., & Thakur, J. S. (2020). Evaluating performance of different grammar checking tools. *International Journal of Advanced Trends in Computer Science and Engineering, 9*(2). https://doi.org/10.30534/ijatcse/2020/201922020.

Shum, S. B., Knight, S., McNamara, D., Allen, L., Bektik, D., & Crossley, S. (2016, April). Critical perspectives on writing analytics. In *Proceedings of the sixth international conference on learning analytics & knowledge* (pp. 481–483). https://doi. org/10.1145/2883851.2883854.

Singh, M. K. M. (2019). International graduate students' academic writing practices in Malaysia: Challenges and solutions. *Journal of International Students, 5*(1), 12–22.

Sirbu, M. D., Botarleanu, R. M., Dascalu, M., Crossley, S. A., & Trausan-Matu, S. (2018, September). ReadME—Enhancing automated writing evaluation. In *International conference on Artificial Intelligence: Methodology, systems, and applications* (pp. 281–285). Bulgaria: Springer.

Soni, M., & Thakur, J. S. (2018). *A systematic review of automated grammar checking in English language.* Available at: https://arxiv.org/pdf/1804.00540.pdf.

Strobl, C., Ailhaud, E., Benetos, K., Devitt, A., Kruse, O., Proske, A., & Rapp, C. (2019). Digital support for academic writing: A review of technologies and pedagogies. *Computers & Education, 131*, 33–48. https://doi.org/10.1016/j.compedu.2018.12.005.

Syafi'i, A. (2020). Grammarly: An online EFL writing companion. *ELTICS: Journal of English Language Teaching and English Linguistics, 5*(2).

Tansomboon, C., Gerard, L. F., Vitale, J. M., & Linn, M. C. (2017). Designing automated guidance to promote productive revision of science explanations. *International Journal of Artificial Intelligence in Education, 27*(4), 729–757.

Ullmann, T. D. (2019). Automated analysis of reflection in writing: Validating machine learning approaches. *International Journal of Artificial Intelligence in Education, 29*(2), 217–257. https://doi.org/10.1007/s40593-019-00174-2.

Wang, Z. (2020). Computer-assisted EFL writing and evaluations based on Artificial Intelligence: A case from a college reading and writing course. *Library Hi Tech*. https://doi.org/10.1108/LHT-05-2020-0113.

Zawacki-Richter, O., Marín, V. I., Bond, M., & Gouverneur, F. (2019). Systematic review of research on Artificial Intelligence applications in higher education—where are the educators? *International Journal of Educational Technology in Higher Education, 16*(1), 1–27.

Zhang, Z. (2020). Engaging with automated writing evaluation (AWE) feedback on L2 writing: Student perceptions and revisions. *Assessing Writing, 43*(10).

10 Intelligent Interlocutors in Teaching Language through Distance Learning Education

Olga I Rudenko-Morgun, Alla L. Arkhangelskaya and Natalia S. Makarova

CONTENTS

10.1 CONTEMPORARY LANGUAGE TEACHING TECHNOLOGIES: ACHIEVEMENTS AND PROBLEMS

The impact of the present pandemic on the development of distance learning (DL) has greatly affected language study. Not only did various problems and contradictions of DL become apparent, but also its great potential. In analyzing what has been lost and gained, we consider what the future holds. This research is devoted to one particular aspect of DL, namely, the use of Artificial Intelligence (AI) in acquiring speaking skills when learning a foreign language in the DL format. The research has been conducted at the Peoples' Friendship University of Russia (RUDN) as part of teaching Russian as a foreign language (RFL) program.

Our review of scientific publications on the topic of DL over the two years of the pandemic (2020, 2021) showed that collective online learning on cloud platforms such as Zoom, Teams and others was established as the main form of DL (Archibald et al., 2019; Kitishat et al., 2020; Iftikhar, 2020; Loranc-Paszylk et al., 2021; Marshall & Ward, 2020; Mohammadi et al., 2021; Tam & El-Azar, 2020; Turchi et al., 2020).

The observed publications make it obvious that full-time education quite successfully converted into a distant format over this period. When considering the problems of online learning, most authors focus on the behaviour of students (Archibald et al., 2019; de Vries, 2021) and this is no coincidence. When working online, it is more difficult for a teacher to manage students' behaviour, motivate their learning activities,

DOI: 10.1201/9781003184157-10

and focus their attention on the topic being studied. More problems arise when it comes to teaching a language, especially at the initial stage, when students at each lesson are required not only to perceive information but also to actively participate in a dialogue, to produce statements in a foreign language that is still not quite familiar to them (Karataş & Tuncer, 2020; Salieva, 2020; Teoh et al., 2016).

Solving the problems mentioned in scientific publications requires a revision of teaching methods, the development of new forms of educational activities, new techniques, and, of course, new teaching aids that correspond to the task at hand.

The authors of this chapter were lucky to have been working in the field of informatization of education for a long time before the pandemic began, having acquired extensive experience in the development of digital multimedia textbooks. So, at the beginning of the pandemic, we were sufficiently prepared for the new learning environment.

Long before the pandemic, it was clear to us that it is extremely difficult to organize communication in a classroom of students who are just starting to learn Russian. To bring students into speech, you need a minimum vocabulary, knowledge of elementary lexical structures, and, of course, word forms, taking into account the fact that the Russian language is inflectional. As a rule, most of the classroom time is occupied by the teacher's explanations and pre-communication training exercises—very little is left for communication itself. Due to all this, from year to year, the "Speaking" test yields the worst results, and methodologists keep searching for effective ways to teach students to speak. Moreover, this happens in the environment of the target language: students of the preparatory faculty of the RUDN University live and study in Moscow.

Back in the 90s, long before the term "blended learning" became firmly established in the teaching methods of various disciplines, we developed electronic resources on Russian as a foreign language. The experience of their use has shown that, equipped with assistance mechanisms and reference materials in the student's first language, they are effective tools for the formation of lexical and grammatical skills in the independent work mode. Needless to say, due to the insufficient development of information and communication technologies, electronic manuals of those years were local and could only support a certain topic in the learning process.

Nowadays, the capabilities of authoring toolkits, such as, Articulate Storyline (https://articulate.com/) and iSpring Suite (www.ispring.ru/), make it possible to create educational and interactive multimedia content: presentations including video clips, animations, drawings, diagrams, simulators, game tasks, and tests. When creating educational materials of this kind, it is important that developers actively use sound. In 2015, after becoming familiar with the features of this e-learning authoring software, we decided to use it to create a learning environment based on blended learning technology.

In blended learning (BL) the teacher effectively combines distance learning with face-to-face instruction on one hand, and innovative training tools with traditional ones on the other. BL technology quickly gained popularity in the modern world of teaching methodology (Stein & Graham, 2020). It is known as the teaching technology of the 21st century. It seems that this idea came almost simultaneously to developers of electronic resources (as a result of their attempts to introduce their innovative means into the educational process, to catch the attention of teachers, including authors of this chapter) and to the authors of DL programs who were not completely satisfied with learning outcomes, especially in the field of learning a foreign language.

Many researchers believe that in the future, BL will lose its special status, and the term "blended learning" will disappear altogether since it will be strongly associated with learning in general (Thorne, 2003; Graham, 2006; Ross & Gage, 2006; Picciano, 2009). The validity of this statement is proven by numerous scientific experiments in this area (Tucker et al., 2016; Fandej, 2012; Medvedeva, 2015).

We were faced with the task of providing conversational practice at the lessons of RFL at the very initial stage of training, using BL technology. Based on our previous experience, we felt that the independent work of students with digital resources should be ahead of the curve. Analysis of publications and familiarity with the flipped classroom model confirmed the validity of our hypothesis.

The *flipped classroom model* (Baker, 2000; Johnson & Marsh, 2016) appeared within the framework of the "blended learning" technology back in the 1990s and did not have a specific name at that time (King, 1993). It assumed the "transfer" of the presentation stage in the field of students' independent work. In the late 1990s and the beginning of the new millennium, studies began to appear that more vividly and clearly described the new model of learning. It was then that the terms "peer instruction" (Mazur, 1997), "inverted classroom" (Maureen et al., 2000), and "the classroom flip" (Baker, 2000) appeared. Nevertheless, the term "flipped classroom", proposed by JV Baker in the early 2000s, has become firmly established in modern science (Baker, 2000).

For modern science, the "flipped classroom" has become a teaching model in which students participate in debates or discussions, having prepared for them at home (mainly through instructional videos). Researches Bergmann and Sams (2012) have remarked that using the "flipped classroom" model helps address the question of how to best utilize classroom learning time. Indeed, the main advantage of this learning model, according to researcher Saxena (2013), is that it frees up class time from routine presentation and memorization of new material and provides an opportunity to use it to organize collective educational experiences for students.

This must have been the exact reason why the "flipped classroom" model was quickly borrowed by the foreign language teaching methodology, which was constantly experiencing a lack of time for free communication between students and teachers in class.

However, we are deeply convinced that the effective use of the "flipped classroom" model implies a radical revision in the organization of the educational process for both teachers and students, and this requires the development of a coherent complex of new teaching aids for students' independent work and collective activities under teachers' guidance.

Before describing the "flipped classroom" model that we propose and the teaching aids that support it, let us turn briefly to the specific reasons for which we chose it for teaching RFL at the initial stage.

The first reason that made us choose the flipped classroom model is that it is relevant for teaching all kinds of disciplines. There is a common trend of a reduction in the number of classroom teaching hours in favor of an increase in the hours of students' independent online work. This tendency is a consequence of the emergence of a qualitatively new information environment: people are eager and able to learn on their own. However, language teachers, working in the traditional "face-to-face" model, find themselves in a difficult position and are forced to change their habitual teaching model.

The second reason, which foreign language teachers worldwide are facing, is different levels of students' language proficiency and their different abilities within the same group. But the question of intensification and individualization of learning for students can also be addressed through self-study, allowing students to fill the gaps, quickly and repeatedly referring to the necessary materials, and work with them in an individual mode.

The third reason stems from the specifics of the Russian language, which for good reason enjoys the reputation of a very complex language. Undoubtedly, Russian (along with some other world languages) has features that learners find difficult: inflection, a specific system of tenses and aspects, indirect word order, free unpredictable stress and more. In learning a language such as Russian, students need to complete a large number of training exercises, which can be long and routine work. It is inappropriate to carry out such work in the classroom, along with the presentation of the educational topic, if there are means that can not only successfully replace the teacher, but even surpass them, using a larger arsenal of possibilities to influence students' perception, attention, and memory.

And finally, the most important reason for us to turn to BL technology and the "flipped classroom" model was, as noted earlier, the acute lack of classroom time for live communication of students with each other and the teacher. It seems to us that this problem, along with other previously mentioned problems, is of current interest to many language teachers worldwide.

We applied the flipped classroom model as we developed a Russian language course (levels A1, A2) for the Digital Preparatory Faculty of the Peoples' Friendship University of Russia (https://langrus.rudn.ru/). We invited students to study each topic through three stages: the preparatory stage, the communicative stage, and the control stage. Figure 10.1 illustrates this.

Let us show the function of the model using the example of work at the A2 level with a digital course (independent work stage) and a course-book (communicative stage).

The preparatory stage in our proposed model takes 30% of the study time. It takes place in the mode of independent work with digital resources. First, students get acquainted with the lexical and grammatical material of the topic, which is presented in two versions:

1. Verbally in English (or in the student's native language) accompanied by figures, diagrams, tables, and so on. In this case, the work of students is controlled; they are given recommendations to carefully study the material, listen to how the examples sound, and repeat them after the speaker.
2. With the help of video fragments, in which the explanation of the topic is accompanied by mini-tasks that activate their attention and control the

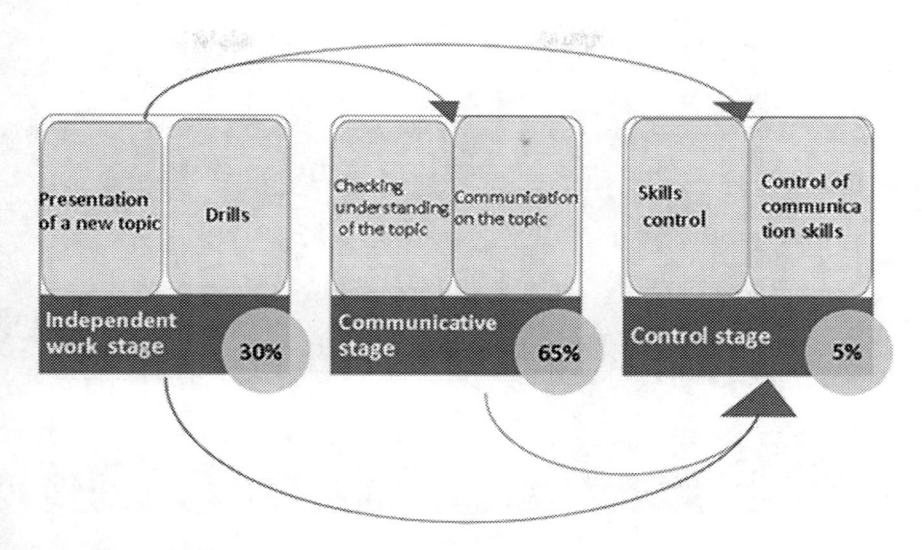

FIGURE 10.1 Diagram Showing the Division of Study Time in the Flipped Classroom Model.

understanding of the presented material. Learners can check the results of their work with the help of pedagogical agents, students of the preparatory faculty of the Peoples' Friendship University of Russia, who perform tasks and comment on them. Students record their answers in the "Workbook", make photos of their work and send them to the teacher. Thus, the second version is preferable since the work of students is, to some extent, controlled.

Figure 10.2 shows video frames sequentially.

Then students perform a set of training tasks, which allow them to fully master the topic on their own. See Figure 10.3.

Note that even if the student does not work very carefully when getting to know the topic, he or she will be able to compensate for the gaps by solving practical problems and turning to help, which can be not only visual but also in audio form.

At the same stage, students are offered texts for listening and reading, accompanied by a system of assignments. The completed assignments are assessed automatically, errors are recorded—the report is available for viewing to the students as well as to teachers.

Final assignments, as a rule, are free written works in the form of detailed answers to questions or essays. This type of work is checked by the teacher.

The second stage takes 65% of the study time. This is teamwork in the classroom. The teacher, firstly, checks how thoroughly the students studied independently: whether they know the vocabulary of the topic, whether they pronounce and perceive new words correctly and use them in grammar patterns. For this purpose, the teacher can use the assignments of the digital course, offering them to students for joint work, as well as a course-book (Rudenko-Morgun et al., 2017) specially created for this

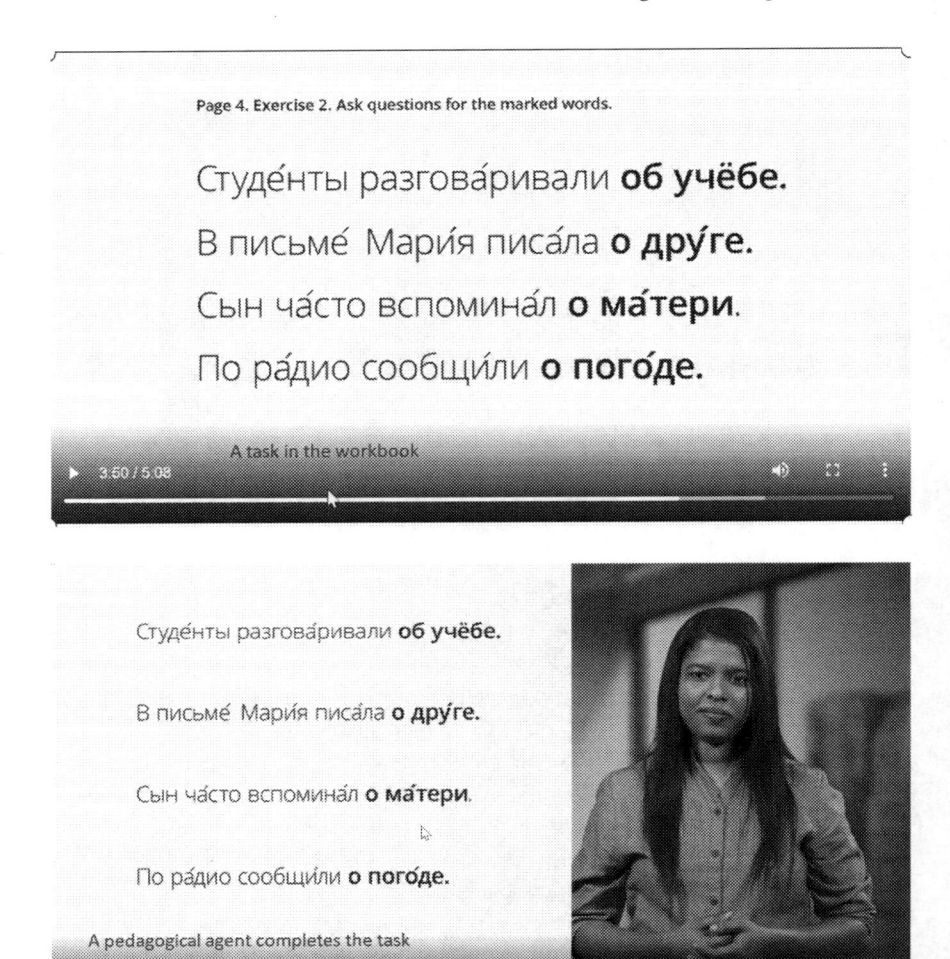

Page 4. Exercise 2. Ask questions for the marked words.

Студе́нты разгова́ривали **об учёбе.**

В письме́ Мари́я писа́ла **о дру́ге.**

Сын ча́сто вспомина́л **о ма́тери.**

По ра́дио сообщи́ли **о пого́де.**

A task in the workbook

3:50 / 5:08

Студе́нты разгова́ривали **об учёбе.**

В письме́ Мари́я писа́ла **о дру́ге.**

Сын ча́сто вспомина́л **о ма́тери.**

По ра́дио сообщи́ли **о пого́де.**

A pedagogical agent completes the task

4:25 / 5:08

FIGURE 10.2   Electronic Course with Grammar Task (on the left) and the Same Task with a Pedagogical Agent Giving her Comments on the Task (on the right).

purpose. Collective face-to-face communication is a great motivation for good independent work—none of the students want to play the role of an outsider in the eyes of his or her classmates. Secondly, and this is the most important, the self-preparation stage allows the teacher to organize a discussion of the stories that the students listened to and read on their own, compelling live communication. Speaking is exactly the kind of work that is given the main learning time.

The control stage is carried out in two formats. During the first phase of control in the mode of independent work, control testing is carried out, which is the final part of each section of the digital resource. It includes the control of reading, listening

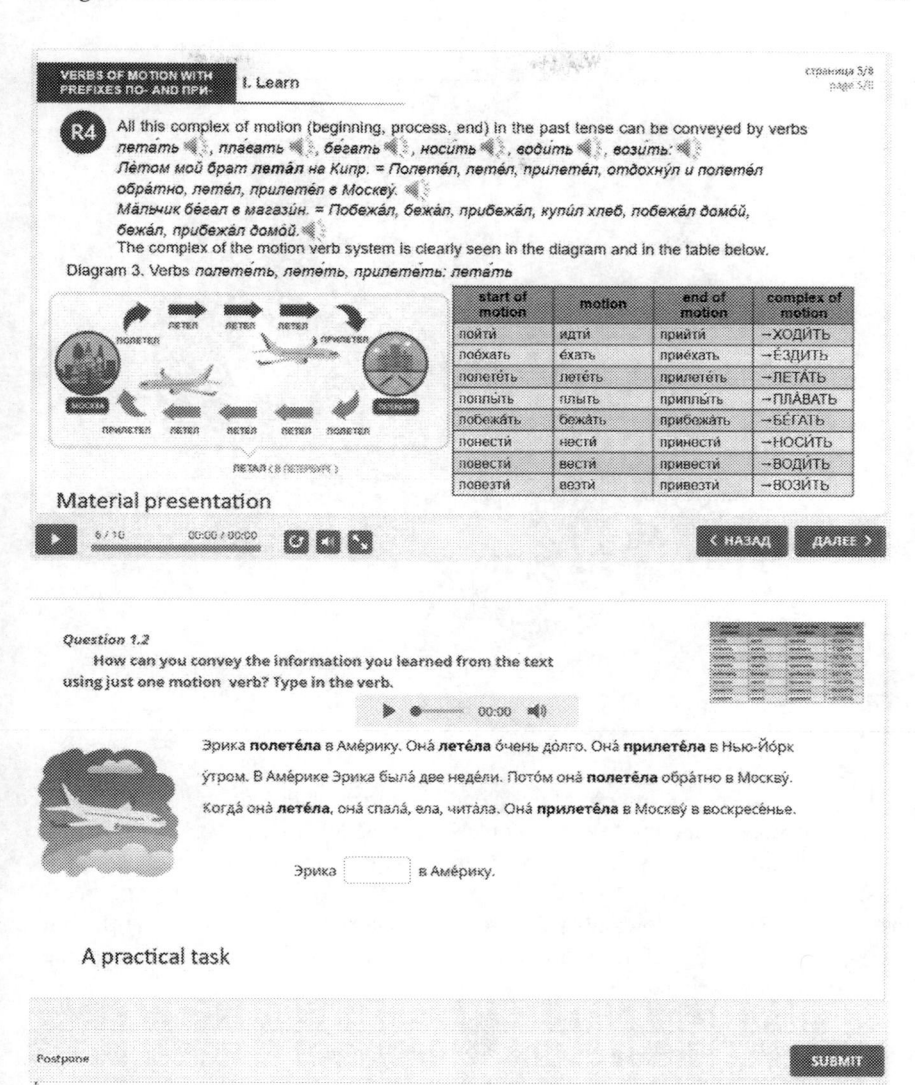

Material presentation

FIGURE 10.3   Grammar Material. Presentation of the Russian Verb of Motion 'to fly'.

(checked automatically) and writing (checked by the teacher). Speaking skills are tested in the second phase of control: the ability to retell the content of the text, answer questions, and participate in a dialogue. This control is carried out by the teacher in the face-to-face mode.

Thus, the developed BL model allows balancing work on all types of speech activity: reading, listening, writing and—especially important—speaking, as it is shown on the Figure 10.5. Figure 10.4 presents independent work, while Figure 10.5, classroom work.

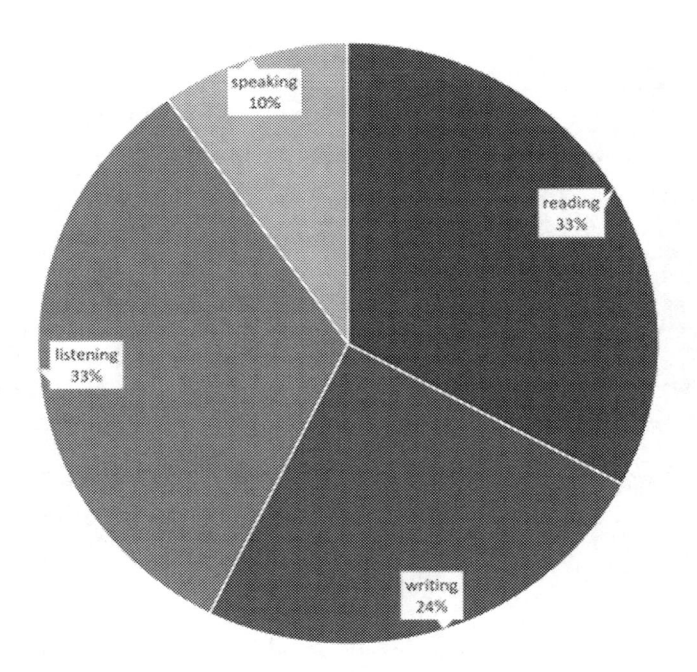

FIGURE 10.4   Independent Work During Blended Learning.

The effectiveness of the presented blended learning methodology was confirmed by an experiment conducted at the Peoples' Friendship University of Russia in 2017.

Control and experimental groups of students were formed according to characteristics relevant for the successful mastering of a new foreign language because they can accelerate or slow down the process of mastering it (especially at the initial stage), and ultimately affect the control results and the purity of the experiment. These characteristics include:

1. The level of language competence (whether a student studied Russian before);
2. Level of education (secondary school, bachelor's degree);
3. Experience in learning a foreign language.

Since the experiment was carried out at the Peoples' Friendship University with students who had just started their studies, it was not difficult to form such groups. After analyzing the students' personal data and the results of entrance testing, two equal groups were formed; all students started learning the language "from scratch", all were approximately equally proficient in English (used as the language of instruction), all had a bachelor's degree and came to Russia to study for a master's degree. The same methodology was applied in the subsequent experiments described in this chapter.

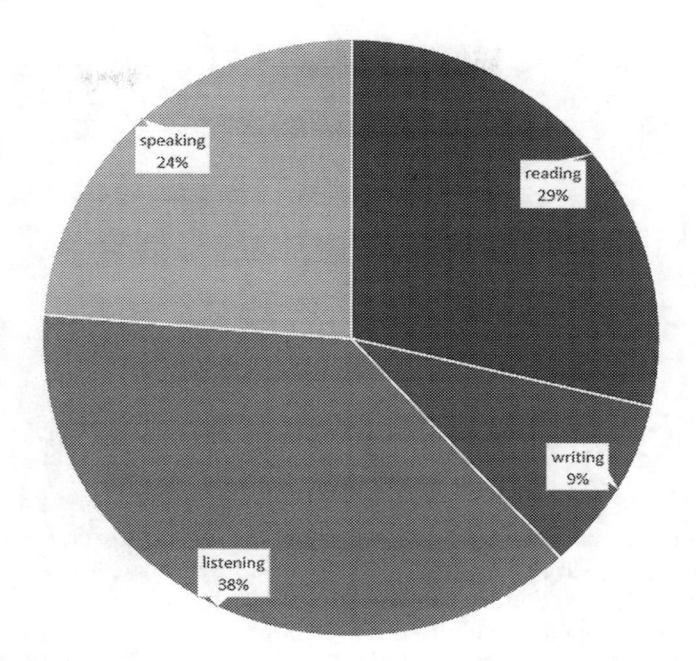

FIGURE 10.5   Classroom Work During Blended Learning.

The experiment involved two multinational groups of foreign students at the Russian Peoples' Friendship University: experimental group (group A) consisted of 11 students and the control group (group B) included 12 students. The total number of students involved in the experiment was 23.

Among the students of both groups were representatives of the Chinese, Vietnamese and Arab nationalities, as well as immigrants from the countries of Africa and Latin America. At the beginning of the experiment, all of them had zero level of Russian language proficiency.

The experiment took place during the first semester, from September to December. Students of group A were taught according to the flipped classroom model described previously, within the framework of BL technology. Training in control group B was carried out in a face-to-face format using traditional teaching aids.

During this period, group A demonstrated higher performance in comparison with group B. This is confirmed by the results of control sessions.

Figure 10.6 shows average test scores at the end of the semester for groups A and B (max. 100 points).

The graph shows that at the beginning of the experiment the results of group A were very low compared with the results of group B. This was due to the difficulties of the organizational period; some of the students did not have the necessary skills for work with digital resources, some of them were not used to learning the language on their own, or they did not participate actively in the collective work in

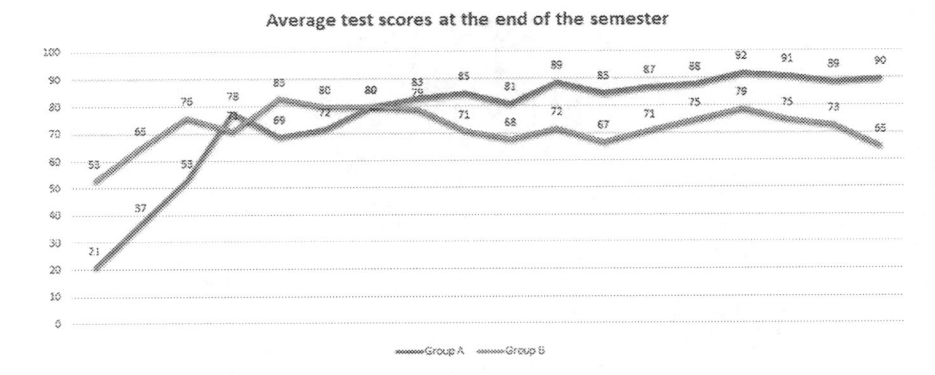

FIGURE 10.6   Average Test Scores at the End of the Semester for Two Groups.

class. But the last control session showed that the students of the experimental group adapted to the new learning conditions, leveled their results with the results of group A students and began to overtake them. This tendency continued until the end of training at the preparatory faculty.

Since we were most interested in the formation of speaking skills, we separately analyzed the results of the final (examination) control of the "Speaking" test (keeping in mind that during the control sessions, competence in all four types of speech activity was assessed). Here are the results of how the students passed the "Speaking" test. Figure 10.7 shows the average score for groups A and B when performing tasks: retelling the text, answering questions and participating in a dialogue.

As you can see in the diagram, the students of the experimental group showed the best results. They demonstrated the ability to communicate fluently on the proposed topic, and they showed freedom in communication with native speakers. Thus, the results of the experiment confirmed our hypothesis that learning according to the proposed model is effective and that this way of learning contributes in particular to the development of oral speech.

10.2   TOWARDS USING AI TO TEACH CONVERSATIONAL RUSSIAN

In the first section of the chapter, we talked about the **flipped classroom** model that we applied and about an experiment that proved its effectiveness in teaching Russian as a foreign language to beginners, in particular for the formation of conversational skills.

However, during the implementation of this model in the learning process, a serious problem came to light: most teachers, whilst using digital resources for students' independent work, still organized their face-to-face lessons the old way. They continued to refer to traditional methods, techniques, textbooks and teaching aids. As a result, the expected benefits of the applied model described so

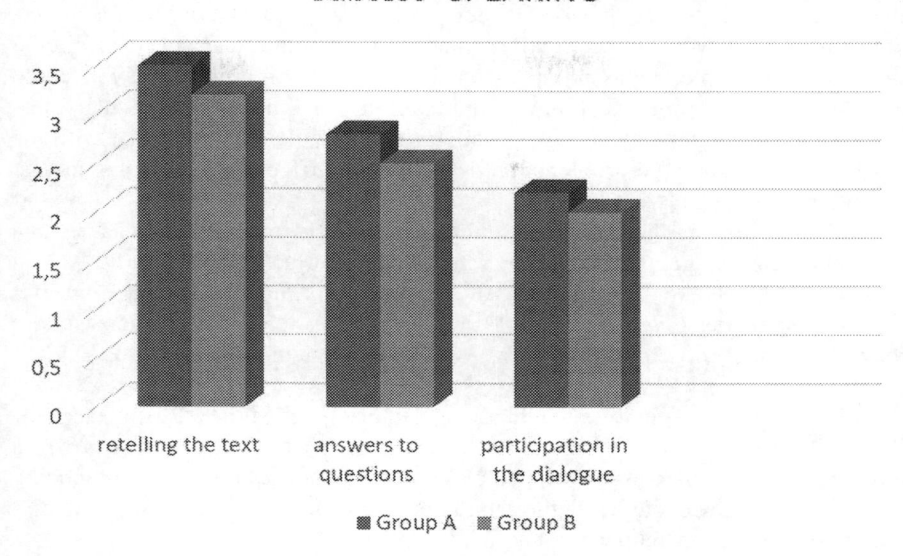

FIGURE 10.7 Average Score for "Speaking" Subtest.

convincingly by researchers Bergmann and Sams (2012) and Saxena (2013) disappeared. According to them, flipped classroom allows efficient use of training time, frees the teacher from having to explain and consolidate educational material in a collective lesson and provides an opportunity to use it to organize "engaging learning experiences for the students" (Saxena, 2013, para 1)—in our case, to organize communication in the target language. Despite the instructions that were offered to the teachers both in writing (in the form of methodological recommendations) and orally (in the form of training seminars), the teachers duplicated the work that the students had already done at home: they explained the educational material, did pre-communication exercises with the students. As a result, there was no time left for speaking practice. Students lost their motivation instead of acquiring it, tired of doing all over again what felt as substantially the same job and not seeing the progress in the development of their target language proficiency. The situation even worsened, compared to traditional education since the study time for face-to-face classes was reduced from 100% to 65% (recall that 30% was allocated for independent work).

We regretfully watched the two basic principles of our model crumble. The first one is the principle of close substantive interconnection and interdependence of all stages of the flipped classroom. It assumes that the teacher builds their lesson based on those competencies that students received during the preparatory (independent) stage of learning. The implementation of this principle is only possible when the teacher is well aware of not only the content of resources for independent work, but also of each student's personal progress in the assimilation of a topic. It is also substantial that during face-to-face class, the teacher uses the educational material closely connected with the content of digital resources, but containing new tasks specifically aimed at oral communication. According to the second principle, a classroom lesson should be as communicatively oriented as possible. All methods, techniques and tasks used by the teacher in such a lesson should stimulate students' free communication with each other and with the teacher. Tasks should be problematic, debatable and affect the scope of students' personal interests.

Without following these fundamental principles, it is impossible to ensure high motivation in the preparatory stage, when students work independently. Also it is impossible to make maximum use of the skills and knowledge gained when it comes to speaking practice in a full-time class. And, finally, it is impossible to move forward in developing language skills and to achieve good results in the control stage.

After reviewing scientific publications (Johnson & Marsh, 2016; Tseng & Walsh, 2016; Ahmed & Asiksoy, 2018), we realized that our situation is far from unique. The researchers of the flipped classroom technology report that practicing teachers do not pay enough attention to the independent work of students with digital resources and do not make use of the results of such work. They continue working in their usual way, without reviewing the content, methods, forms and techniques of their classroom activities (Johnson & Marsh, 2016). Researchers note that teachers underestimate the importance of their interaction with students in class: teachers are often not ready to change the usual methods of organizing classroom activities or do not know how to do so (Tseng & Walsh, 2016). The researchers Christopher and D. March (2016) point out the basic principles of effective flipped classes and say that they began to realize the number of changes that "traditionally" trained language teachers will have to face in their lessons in the classroom (p. 62).

As evidenced by our experience of introducing the flipped classroom technology into the process of teaching RFL at the preparatory faculty of RUDN University, even the presence of special training materials for face-to-face work interconnected with digital resources does not solve the problem. Let's not blame the teachers, as conservatism is a characteristic of their profession: it is dangerous to experiment when you work with young people, and the method that you have used for many years has been quite successful. However, the pandemic has presented teachers with the inevitability of conducting classes online, and so has made an irreversible transition to a new and undoubtedly more effective methodology. It is an ill wind that blows nobody good, as the proverb says.

As has already been mentioned, our many years of research in the field of information computer technology (ICT) application in language teaching contributed to the fact that, by the beginning of the quarantine and the forced transition of the educational process to remote communication with students (April 2020) at the preparatory faculty of RUDN University, the model of the flipped classroom and the complex of means supporting it were not only developed, but also experimentally tested. We had a good reason to assume that, on this basis, it was possible to organize distance learning quickly and with minimal losses. Digital resources and educational materials for independent work started to be used more actively in the educational process. At the same time, new problems arose in the new working environment; in the independent mode as well as in online classes on Zoom and Teams platforms.

As always, the main problem has been speaking, which, even during face-to-face training, causes the greatest difficulties of all language skills (Banditvilai, 2016; Leong & Ahmadi, 2017; Vurdien, 2019; Karataş & Tuncer, 2020). The teacher makes great efforts to enhance students' fluency and help them to overcome psychological barriers that prevent them from speaking. In an online lesson, this is even more difficult: students do not hear (or pretend they do not hear) a question, do not turn on the microphone or turn off the camera for a long time, read the answer instead of speaking without visual reference to the text and so on. Teachers joke: "My lesson in Zoom is like a seance. I keep calling out for my students without seeing them".

An anonymous survey conducted among students revealed the main reasons for their passivity. As we expected, of the four suggested answers to the question "What type of speech activity are you worst prepared for after working with digital resources on your own?", the majority of the respondents chose "speaking". For this other question: "What do you think are the reasons for your passivity in the online lessons?", the main responses were: poor Internet operation ("I have difficulty hearing the teacher's questions and the remarks of my groupmates") and a feeling of insecurity ("I'm afraid to pronounce the words incorrectly", "I feel uncomfortable because my speech is being recorded").

An analysis of the video recordings of the educational process showed that the proportion of work on certain language skills during classroom work in the distance learning format has changed compared to blended learning. Figure 10.5 demonstrates this. Let us compare these results with the results in Figure 10.8 from section one. We can see that the share of listening has increased significantly, which is not surprising: the teacher is forced to repeat the questions and the students' names several times, they comment on the students' answers more often, as they find it difficult to grasp the students' reaction and assess how well they understand the topic under discussion. The share of reading also increased because, often, after not receiving an answer to their question, the teacher takes the students back to the text or the dialogue and asks them to read the text again in order to solve the problem.

As a result of a comprehensive study of the problem, we assumed that the tools we were using to train speaking at the stage of independent work were not effective enough. We made a list of such tools and considered each one separately. Our list included

- general recommendations and commands preceding the task, such as "repeat", "pronounce", "speak";
- tasks such as: say and type the missing words/sentences of the text, replicas of the dialogue;
- Figure 10.9 shows a screenshot of a video clip demonstrating the correct pronunciation of sounds (hard and soft consonants), words, phrases;

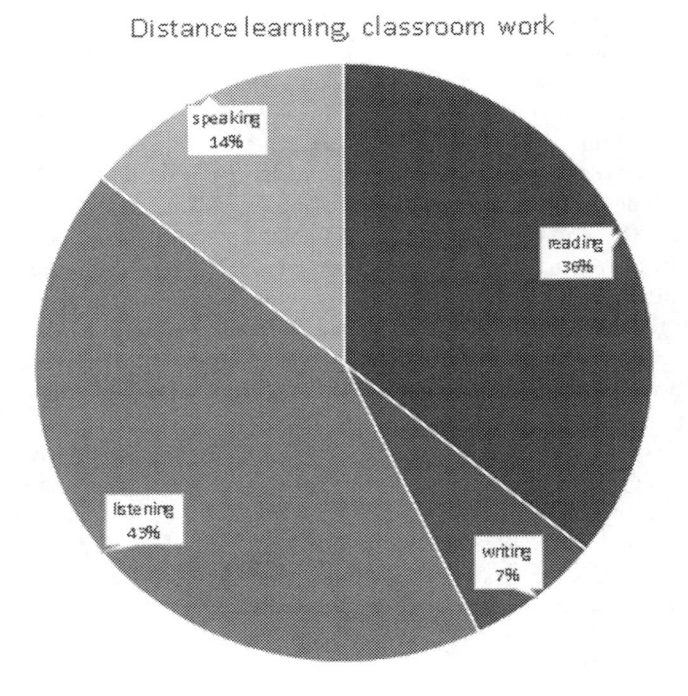

FIGURE 10.8   Classroom Work in the Model of Blended Learning.

• animated scheme of the vocal apparatus, clearly demonstrating the articulation of the sounds, as presented in Figure 10.10 to follow.

A survey conducted among the students (they answered the questions "Do you repeat words and sentences after the speaker when you work with the course?" and "When you complete the tasks, do you say your answers out loud or just to yourself?") revealed the reason for the inefficiency of our tools. The most common response was "Yes, I repeat/pronounce words and sentences, but I'm not sure if I'm doing it correctly".

Thus, the reason was found: language skills such as reading, listening, writing were controlled automatically and quickly but the tools for developing speaking skills, due to the limited capabilities of Articulate Storyline (https://articulate.com/) and iSpring Suit (www.ispring.ru/), only had functions of demonstrating (video fragments demonstrating articulation, animations of vocal apparatus) and stimulating ("listen and repeat", "restore the missing information, say, then type" commands). In fact, all of them developed listening skills rather than speaking skills. Undoubtedly, listening affects the mechanisms of speaking and contributes to its improvement, but this is not enough. No matter how seriously students work during the self-preparation stage, encouraging them to speak is not an easy task, as our practice has shown. Imagine that a student completed all the tasks thoroughly: not only did they listen,

FIGURE 10.9   Classroom Work in Distance Learning Format.

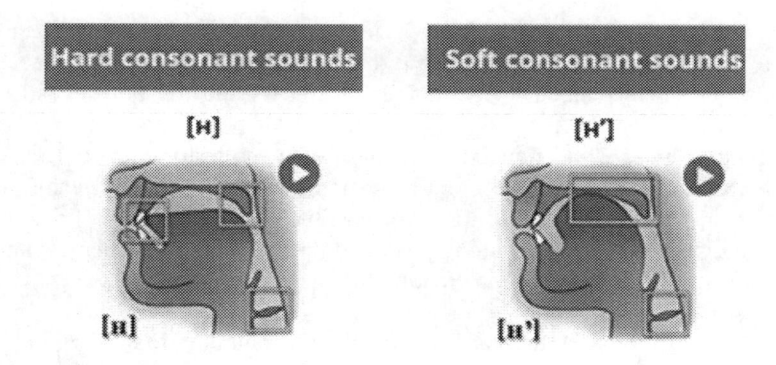

FIGURE 10.10   Video Clip from an Electronic Phonetic Course.

but they also repeated after the speaker; not only typed in the answer but also pronounced it. But finding themselves in a situation where one must speak in front of the audience and in front of the camera, they feel insecure, because during the preparation for communication they received no feedback in the form of assessment and correction.

Thus, we came to this conclusion: it is necessary to improve our tools or propose new ones that allow you to train speaking skills at the self-preparation stage. Since the improvement of the tools (described previously) did not seem feasible, we turned to the possibilities of AI. Namely, to the tools of speech recognition (Bhatt et al., 2020; Buddhima & Keerthiwansha, 2018; Chen et al., 2020; Dodigovic, 2007; Lutskovskaia et al., 2019; Metatla et al., 2019). The analysis of publications shows interest on the part of the researchers, who deal with the problems of learning foreign languages (Tregubov, 2020).

It should be said that we have already had experience with speech recognizers. In 2007, in cooperation with the programmers from the Russian company IstraSoft (www.istrasoft.ru), we created an electronic course "Russian language with a computer. Step I" (https://istrasoft.ru/ru/programmy/russkij-jazyk-s-kompjuterom-shag-1.html), which was geared towards foreign students starting to learn Russian. It included the full educational material of the introductory phonetic course, which plays an important role in teaching foreign students. At the initial stage of their learning, they get acquainted with the phonetic system of the Russian language, the articulation of Russian sounds, the characteristics of stress and its implementation in various types of words and the main types of intonation patterns. "Russian language with a computer. Step I" was our first attempt to implement the flipped classroom model in the design of a course. It was assumed that the learners work with the course independently: all supporting and reference information was presented at the user's choice of five languages: English, French, Spanish, German and Chinese. Whilst working on the course, learners obtained the necessary knowledge of Russian phonetics and graphics, practiced pronouncing Russian sounds, words, phrases, and then consolidated and corrected their skills in the classroom under the guidance of a teacher. Speech recognition technology was applied in the course to support the independent work of learners.

It is important that the IstraSoft programmers, in accordance with the task set by the teaching methodologists, achieved the division of the word into elementary segments that corresponded to the sounds of speech and did not depend either on the speaker or on the language. The extraction of phonemes from the stream of continuous speech, their encoding, and subsequent restoration took place in real time. Previously, speech recognition systems were not able to perform such segmentation—the smallest unit for them was the word, so it was impossible to work on the pronunciation of individual sounds. Processing of speech at the phoneme level was a step towards the creation of a new generation of speech recognition systems.

The electronic course "Russian language with a computer. Step I" was actively used at the preparatory faculty of RUDN University. At that time, most of the students did not have laptops, and the work took place in computer labs. This allowed to monitor their work and spot problems. The most important problem was the control of pronunciation. Pronunciation control was carried out in the course in two different ways. The first option was a comparison of oscillograms. See Figure 10.11 to follow.

Since the comparison of oscillograms seemed as a long and tedious job, not suitable for everyone, we asked the developers to implement a quantitative, ten-point articulation assessment. You can see this evaluation in Figure 10.12.

Although a quantitative assessment greatly facilitated and intensified the students' work, this solution did not completely satisfy us: we wanted to speed up the control process even more to achieve a greater degree of visualization. Of course, the working conditions of students affected the situation a lot; as we already mentioned, their work took place in a computer lab and was limited by a certain time frame. Students worked at different paces as some of them used oscillograms, while others turned to quantitative assessment; some tried to

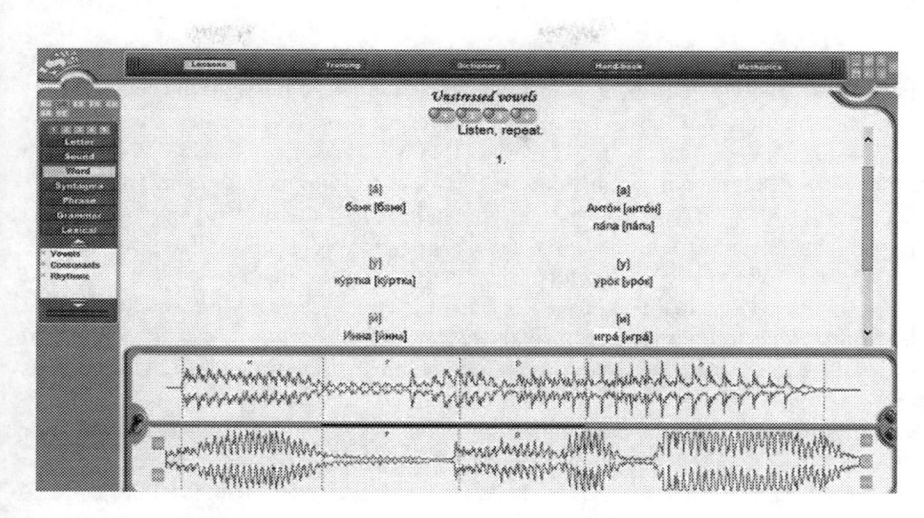

FIGURE 10.11   Articulatory Apparatus Animation.

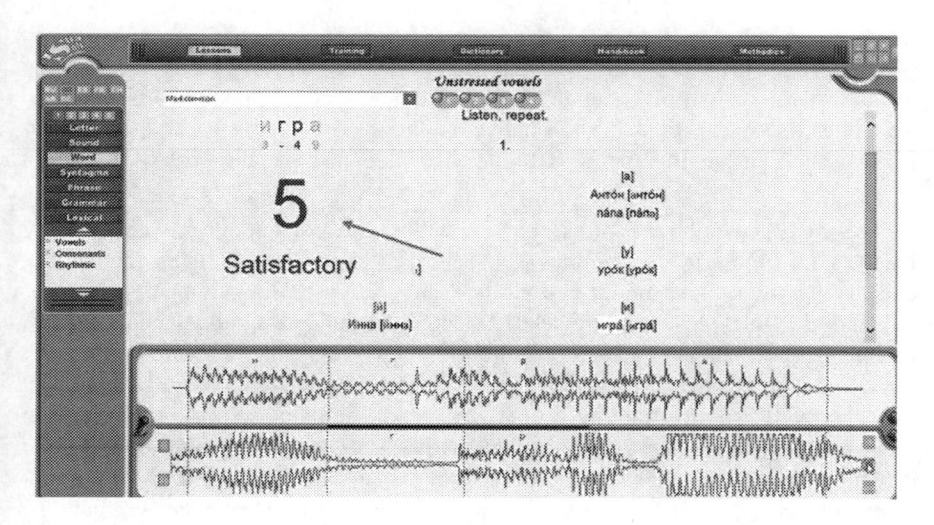

FIGURE 10.12   A Screen Shot from the Electronic Course "Russian language with a computer. Step I".

achieve the maximum grade, and others were satisfied with the average score. As a result, the difference in the volume of the material studied and the quality of its assimilation was too great, and this made it difficult for the teacher to work in the face-to-face format.

In 2020, with the help of the authoring software Articulate Storyline (https://articulate.com/360/storyline), we developed a new introductory phonetic course specifically for the Digital Preparatory Faculty of RUDN University (https://langrus.rudn.ru/local/crw/course.php?id=6).

It differed from the one described earlier not only in its modern interface, but also in content, since the capabilities of this software made it possible to include video clips, animations, and make training tasks more diverse and exciting. But the software itself did not allow creating exercises that train articulatory skills. Web Speech API (https://elearningdifferently.com/speech-recognition-and-storyline/) helped us accomplish this task as it allowed us to include voice data in web applications and implement speech recognition in Articulate Storyline 360 (Suhov, 2013).

When developing scenarios for the formation of articulatory skills, we focused on the characteristic features of the Russian phonetic system, taking into account the difficulties that foreign learners usually face when mastering it. At the very initial stage of language learning, work on articulation most often turned into a monotonous routine. Whilst designing activities for the "Speak!" lessons of the new course, we wanted students to not only do training work, but also, as far as possible, expand their knowledge of the target language. For example, whilst working on the articulation of vowel sounds [a], [e], [o], [y], students get acquainted with Russian interjections. Short words such as "oh", "ah", "uh", "eh", as well as onomatopoeic words, are perfectly suitable for accomplishing this task. The skills acquired during independent work are then consolidated in "face-to-face" classes: the teacher invites students to voice pictures or video fragments, expressing various emotions in Russian. Based on this material, a teacher can easily organize a dialogue of cultures (with students who are only familiar with five Russian vowels and a few consonants!), asking them to reproduce the interjections of their native language. This kind of work is especially interesting for international groups. Figure 10.13 demonstrates the work with Russian interjections.

As we can see in these Figures, working with a task like that, the student instantly receives a response to their actions in the form of an inscription under the picture, and by clicking on the Submit button, they can see a verbal assessment of his pronunciation. If they fail, they are asked to complete the task again. Needless to say, this algorithm for completing a task requires much less time compared to similar work with the electronic course "Russian language with a computer. Step I", that we described earlier. Additionally, a student at any time can receive help either in the form of a grammar reference in their native language, which will remind them of what position his organs of speech should take for pronunciation of this or that particular sound, or in the form of an animated diagram of the speech apparatus. Such conditions are comfortable for the student; they are not ashamed of their errors as neither their teacher nor their group mates are listening, and as a result, they feel more confident in the subsequent face-to-face lesson.

A great difficulty for foreign students is the presence of voiced and voiceless paired consonants in the phonetic system of the Russian language. In traditional teaching practice, the teacher spends a lot of time setting up and practicing their pronunciation. As a rule, the teacher does not have enough time for individual work—hence the numerous articulatory errors made by the students. Our digital resource enables each student to work in an individual mode, first completing the listening and then the speaking tasks. This type of task is shown in Figure 10.14.

These screenshots show that we are faithful to the concept we have chosen—when forming articulatory skills, we are constantly relying on meaning. We are deeply

FIGURE 10.13 A Screen Shot from the Electronic Course "Russian language with a computer. Step I" with Evaluation.

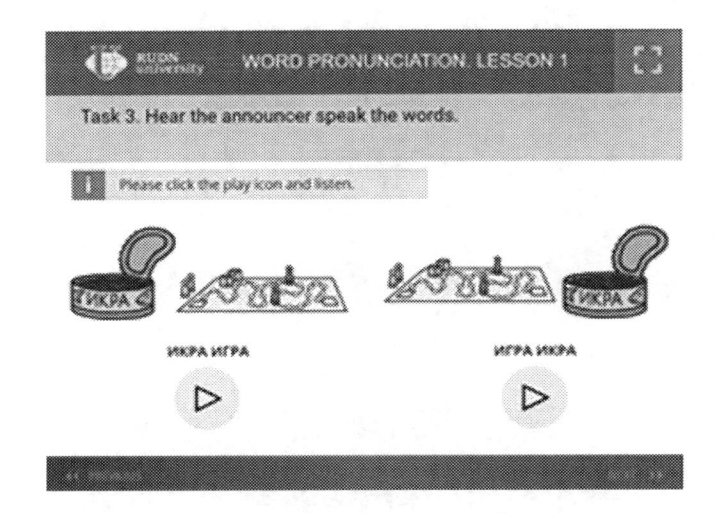

FIGURE 10.14   The Work with Russian Interjections in the Electronic Phonetic Course.

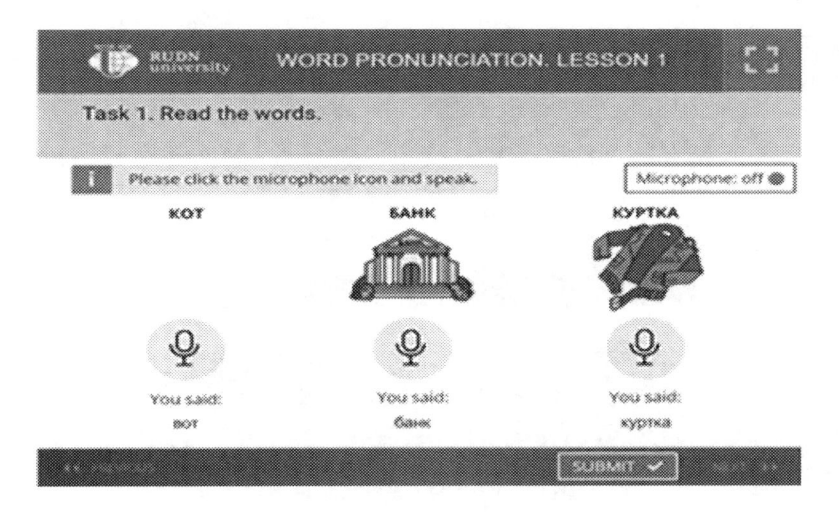

FIGURE 10.15   Screen Shot from the Electronic Phonetic Course with Written Russian Words Similar in Pronunciation.

convinced that semantic differences help the students to comprehend the significance of the phonetic phenomenon and its relevance to distinguishing the meaning, which motivates the students to constantly work on their pronunciation.

Turning to the speech recognition tool allowed us to improve the quality of students' work with the lexical material of the course. From now on students learned new vocabulary not only by listening to and typing new words, but also by working on their pronunciation. See Figure 10.15.

We were able to correctly conduct an experimental test of the impact made on our students by the tasks created with the help of a speech recognition tool. We made use of the fact that the study groups of the Digital Preparatory Faculty in 2020 were formed at different times. The control group (group B) started classes at the beginning of September when the tasks described earlier had not yet been included in the Introductory Phonetic Course. The experimental group (group A) was formed and began to work with the course a few months later, when the "Speak!" lessons had already been added. The groups were formed in accordance with the principles described earlier in section one. Having analyzed the face-to-face lessons recordings of both control and experimental groups, we realized that the time of students' speaking in online lessons with a teacher is significantly larger in group A, thanks to the preparatory independent work on the pronunciation. Figures 10.16 and 10.17 show this.

The control test, conducted after the students passed the introductory phonetic course, showed that while demonstrating their reading and speaking skills, students

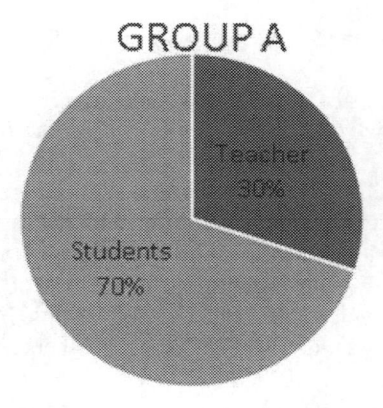

FIGURE 10.16 Screen Shots from the Electronic Phonetic Course. Recording Pronunciation.

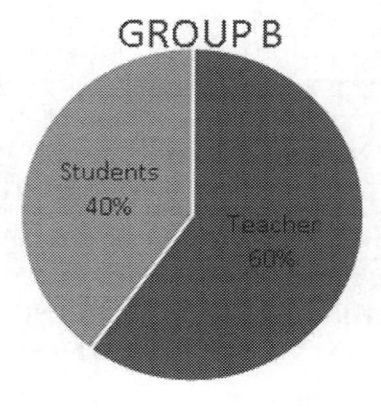

FIGURE 10.17 Result of Experimental Test with the Help of a Speech Recognition Tool in Experimental Group.

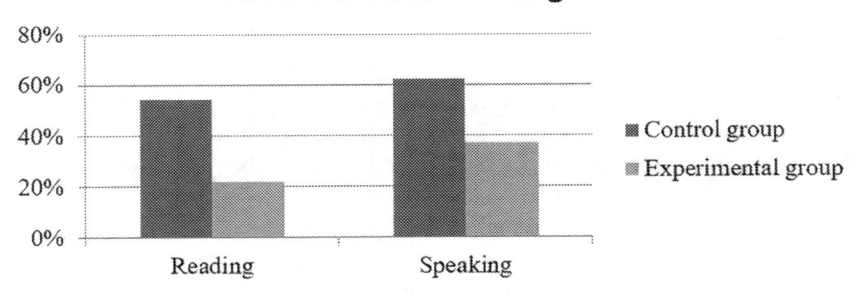

FIGURE 10.18   Result of Experimental Test with the Help of a Speech Recognition Tool in Control Group.

of the experimental group made fewer mistakes than students of the control group as shown in Figure 10.18.

In addition, the students of group A showed a more solid mastery of the vocabulary of the course, which was clearly facilitated by the tasks where sounding of pictures were involved.

At the same time, testing the tasks with the speech recognition function and questioning students revealed some important problems. It turned out that the tasks for pronouncing words—which radically differ in the set of sounds—are quite correct, easy to understand, and do not cause any difficulties. But the tasks aimed at distinguishing between paired voiced and voiceless consonants ([ikra]—[igra]), as well as paired hard and soft consonants ([byl]—[byl']) and others similar in their characteristics ([hlam]—[hram]), cause difficulties and require an unnaturally emphasized articulation. Thus, more advanced instruments are needed to recognize sounds similar in articulation, to identify errors typical of foreign learners and to correct them.

Having studied this issue, we discovered that such tools already exist and are successfully used in Russian medicine in restoring the speech of patients after a heart attack. Based on the Sphinx search engine (https://forex-discount-store.com/download-sphinx-eg-ea/), an AI was created that can clearly identify the set of sounds that a person uttered, and then report their characteristic errors. All the words that a patient pronounces incorrectly are listed for further work on speech defects (Bulanov, 2019), (Zawacki-Richter, O., et al., 2019). Undoubtedly, such an instrument would be very helpful for learning and teaching foreign languages. We consider the task of its implementation in the field of teaching phonetics promising for further research.

10.3   INTELLECTUAL INTERLOCUTORS IN THE DISTANCE LEARNING ENVIRONMENT

Undoubtedly, the task of teaching the articulation of sounds, their reduction and intonation is key at the very initial stage of learning the Russian language. This is the foundation without which it is impossible to continue learning it. In the second section of our chapter we talked about how, in the context of distance learning, it was possible, by turning to available tools, to approach the problem of forming the foundations of speaking using AI. When starting to develop A1-level content in the DL format, we already understood perfectly well that the development of these skills would again become a stumbling block for students studying a language away from a Russian speaking environment, given the limited time of communication with their teacher, the native speaker of their target language.

Within the BL framework, we successfully used the plot-based introductory speech course "Russian with Russian friends" (https://langrus.rudn.ru/local/ crw/ course.php?id=18) that we developed earlier. Each section of the course was aimed at teaching all four language skills, among which most attention (and importance) was given to speaking. The plot of each lesson of the course develops first in reading materials (Lesson "Reading"), then in texts for listening (Lesson "Listening"), and, finally, ends in dialogues (Lesson "We Talk").

We would like to draw your attention to the plot-based organization of the course since it was the reason that ultimately led us to the need to use AI. It must be said that the plot plays an important role in creating the conditions for communication. The plot helps the teacher rely on natural life situations when organizing a dialogue with students, if such situations are presented in the didactic material. The need to teach on the basis of situations is recognized by all teachers, but they understand it differently. Descriptions of situations (such as, "Imagine, you are late for class. What would you say when you enter the classroom?"), which are often used by authors of the course-books, usually "do not work" because they are perceived as a purely academic task. They leave students indifferent nor do they encourage conversations, discussions and therefore cannot develop speaking skills. The willingness to participate in communication appears only in a real or reconstructed situation that affects the interests of the speakers (Passov, 1991). It is exactly such situations that a plot-organized course can provide to the teacher and students, of course, if the plots of its lessons are natural and, most importantly, related to problems and topics that are relevant for students. The plot-based organization of the course "Russian with Russian Friends" excludes a conventional approach to working with educational material. Moreover, when communication tasks are based on real-life situations, students become aware of the close connection between grammatical and semantic meanings and understand that knowledge of grammar is necessary if they want to understand their interlocutors and be understood by them.

As we have already said, the course pays a lot of attention to work with the dialogues of the "Speak!" lessons. Students are not encouraged to learn the dialogues by heart—they memorize them through the process of completing exciting and useful tasks that develop listening skills, contribute to vocabulary retention and speech patterns as well as the mastering of Russian intonation. Here are the tasks that students are asked to complete:

- listen to the remark, repeat it, and then, from two words that sound similar, choose the one that you have heard in the remark;
- listen to the remark, repeat it, type in a punctuation mark that needs to be put at the end:

- listen to the remark, repeat it, select the character to whom it belongs;
- listen to Marina's remarks from the dialogue, during the pauses say the missing remarks of Ivan (Marina and Ivan are the characters of the course);
- play out a dialogue between Marina and Ivan with your group mates.

Figures 10.19, 10.20 illustrate the last two tasks. These tasks will be discussed to follow.

Unlike the previous ones that are supposed to be completed at home, digital courses are performed "face-to-face", as only a teacher can assess the quality of their implementation. In the conditions of the BL model, this suited us perfectly. The survey showed that teachers had enough time in the face-to-face lessons to check the effectiveness of working with a digital resource using the last two assignments, and to organize a role-playing game based on a plot that is similar to the plot of the lesson. Let us remind you that the time ratio of independent and face-to-face work in BL was 30% to 70%. The change in the distribution of time in DL (70% to 30%), as well as the conditions of work in the online format, led to the consequences that we already described in section two: live communication faded into the background, playing out dialogues was most often replaced by reading them. The introductory speaking course no longer fully corresponded to the role to which it was assigned.

Since the reason was already clear (it was the impossibility of practicing speaking skills during independent work), we decided to turn to the same speech recognition tool (Web Speech API) that we had used in the development of the introductory phonetic course. As a result, the set of tasks was supplemented with tasks of the type: "Read

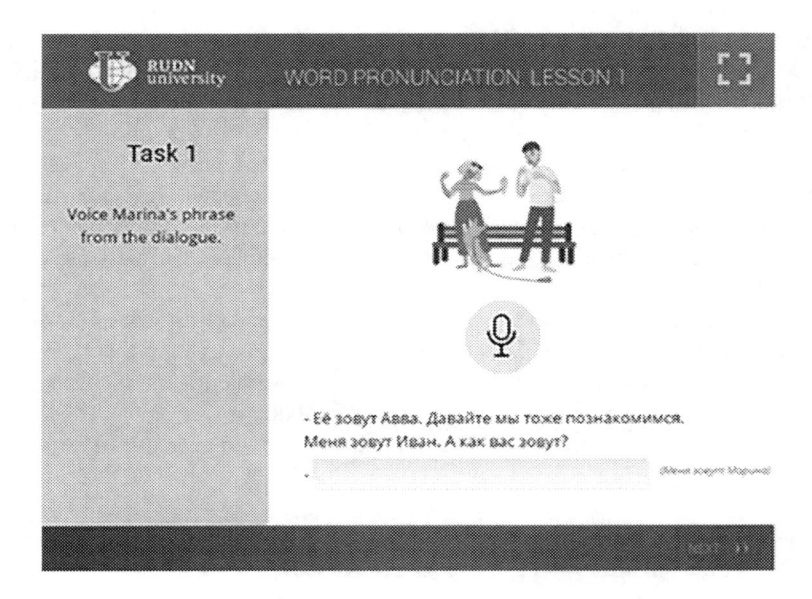

FIGURE 10.19 Mistakes Made During Control Test, Conducted After the Students Passed the Introductory Phonetic Course.

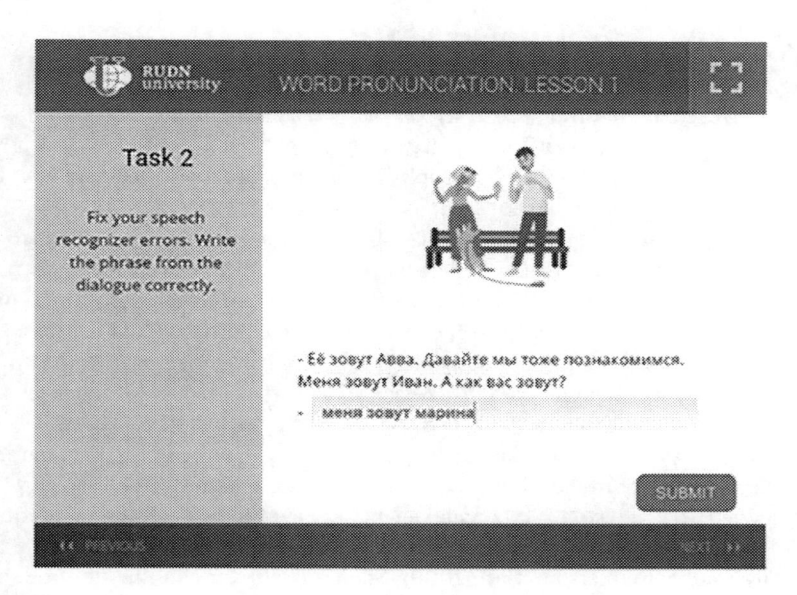

FIGURE 10.20 Electronic Speech Course Screen 1 (Dialogue replicas with the empty input field).

Ivan's/Marina's remark from the dialogue and answer it". But when adding these tasks to the course, we encountered the following problem: Web Speech API quickly recognized the speech and recorded it, but completely ignored capital letters and punctuation marks. And for us, this function was fundamentally important, since in our digital course we organized work based on intonation and its reflection in writing.

Our many years of experience in the development of digital resources taught us to adapt to the capabilities of various platforms and editors, and the solution to the problem was found quite easily. We developed a new kind of task, to be completed in two steps.

In the first step, the student voiced a line from a dialogue. If a student did it correctly, in the second step they were asked to correct the errors in the text displayed on the screen: replace (where necessary) lowercase letters with capital ones and put punctuation marks in the right places. Note that as a result of overcoming such problems, the tasks often do not lose, but, on the contrary, gain methodological value. And so, it happened in this case: the students not only improved their speaking skills but also learned to spot errors in the written text and to correct them.

Supplementing the set of tasks in the course with remarks from the dialogue solved an important problem: the students, as the experiment showed, firmly memorized and retained speech patterns important for communication and actively used them. These patterns have become an excellent mnemonic model for generating similar utterances. When the students, who were already in the second semester, answered the question: "Which of your first textbooks do you most often remember?" and were offered a choice between all the educational material they used at the beginning of their studies, the majority chose the digital course "Russian with Russian friends". Here's what they wrote: "When I need to tell you where I am studying, I just remember Ivan's phrase from this course: 'I study at the Faculty of History' ", "I am sure I will never make a mistake answering this question. Thanks to this course, I am not afraid to speak Russian", "I found it interesting to follow the story about the characters of this course. They taught me what to say and how to behave in different situations. For example, I know what to give a Russian friend for a birthday and how to arrange a meeting with him".

Thus, we can see that the characters of the course have become pedagogical agents, teaching the students how to behave and communicate with Russians in different everyday situations. They have indeed, but not completely. We wanted to use the capabilities of AI not only for students to memorize replicas from the dialogues; we wanted them to learn to communicate freely on given topics whilst working on their own. This, we thought, would increase their activity in the online classroom, in conversations and discussions and during role-playing games. We also wanted students not only to observe what was happening to the characters of the course but to be able to take part in their life and to communicate with them.

Let's consider the term "pedagogical agent". It entered the teaching methodology and practice as a computer-animated character, functioning in the interactive educational environment (Johnson et al., 2000). Such a character can play the role of an instructor or navigator for users, and also, in recent years, give them knowledge in different subjects. The history of the pedagogical agent as a phenomenon is inextricably linked with the appearance and development of the first computer animation when, in 1960, Ivan Sutherland offered the world the technology of creating drawings directly on the monitor screen (Oakes, 2007).

It is interesting that Russian teachers, including one of the authors of this chapter, were the first to draw attention to the educational potential of these animated characters. Back in the early 1990s, before the era of multimedia, RFL teachers together with the students of the Faculty of Computational Mathematics and Cybernetics at the M.V. Lomonosov Moscow State University created a computer linguistic educational game for learning RFL named "The Case Detective" (Vasilyeva et al., 1991). The animation was widely used in this game. There was a multi-coloured cube rotating to illustrate the meanings of cases, a parrot that tore a mysterious note found in the house of a Russian emigrant, words on the screen changed their shape, nouns' inflections flew out of a pistol barrel and detectives exchanged encrypted notes. The first pedagogical agent, endowed with educational functions, acted in this game. His name was Dr. Grammer: he set educational tasks for the students, helped to solve them and controlled their actions. So, for example, when the number of user errors exceeded the permissible level, Dr. Grammer's permanent attribute—a pipe—fell out of his mouth. Figure 10.21 introduces this character to you. Surprisingly, the programme is still running.

This experience, although successful enough and approved by domestic and foreign experts, could not get widespread before the advent of multimedia technology— drawings and animations took up too much memory in computers of that generation.

The most productive years in the development of pedagogical agents were 1998– 2000. During this time, several American universities at once developed teaching agents. All of them were human-like or cartoon characters that could perform simple tasks: instructing users by telling them or demonstrating what to do.

FIGURE 10.21 Electronic Speech Course Screen 2 (Dialogue replicas with the completed input field).

For example, the Information Sciences Institute at the University of Southern California has developed two pedagogical agents: Steve (Soar Training Expert for Virtual Environment) and Adele (Agent for Distance Learning: Light Edition) (Johnson et al., 2000, 48–49). Steve was introduced in 1998 as a 3D animated character functioning in an interactive training programme for future naval engineers (the agent helped them complete the tasks using a computer model of a ship's engine). Introduced in 1999, Adele was a simpler 2D educational agent, embedded in the web browser interface of medical students to integrate Internet content with teaching materials.

North Carolina State University developed and introduced three teaching agents at once in 1999 (Johnson et al., 2000, 50–52) Herman-the-Bug, COSMO and WIZLOW. The first one was introduced into an interactive educational programme created for botany students. The other two agents (COSMO and VISLO) were developed for computer science students. Educational programs with pedagogical agents included in their interface quickly proved their effectiveness (Johnson et al., 2000)

Among the additional advantages of pedagogical agents, researchers point out the increased interest in the educational material and hence the increased motivation of students, their growing confidence in the knowledge gained due to the elimination of a subjective attitude of the student to the teacher, and an opportunity for a real teacher to transfer the functions of a controller to virtual agents, playing himself the role of an assistant and friend, in accordance with modern trends in teaching methodology.

And that is not all. For those who learn a foreign language, pedagogical agents can become models of ethical and speech behaviour and "living" carriers of the sociocultural characteristics of the target language country. Our own experience in digital resource development has only reinforced this view. In 2006, the Russian company "1C" presented a line of electronic textbooks on the Russian language for schoolchildren, edited by Rudenko-Morgun (Rudenko-Morgun & Arkhangelskaya, 2006, Rudenko-Morgun et al., 2015; Rudenko-Morgun & Arkhangelskaya, 2016).

The basis of the hypermedia textbook was made up of virtual lessons, where a group of pedagogical agents functioned and interacted with each other and with real users—they were animated cartoon characters with different personalities. See Figure 10.22.

As you can see in the figure, the screen is similar to the Zoom or Teams screens that are so familiar to us since the beginning of the pandemic. On the left, there is a virtual class: teacher Andrei Ivanovich and three students Vasya, Liza and Anfisa. The multimedia lesson can be studied independently, or a teacher can use it whilst working with the class in a face-to-face classroom or in an online lesson. Pedagogical agents facilitate the inclusion of real students in the discussion of the topic, assign them problematic tasks and comment on their mistakes made when performing interactive tasks (collectively or individually on an interactive board). They argue with the "teacher", ask them interesting questions, play linguistic and role-playing games with real students, and create mini-projects. As can be seen, the role of the pedagogical agents is not limited to the role of an instructor, assistant or source of reference information. They show an example of behaviour in the educational team; they activate, liberate students, awaken their creative imagination and motivate them to study the subject.

Children worked with the electronic textbook both in class and at home. The virtual classroom evoked the most positive emotions among schoolchildren, teachers and parents. They wrote to the developers: "I like Vasya very much, he is so cool—he comes up with something interesting all the time", "My students have become more active, they express their opinion more freely, trying to imitate Vasya, Liza and Anfisa. They talk more in the lesson", "My son has always been very uncommunicative: he was afraid to ask questions and speak. But now I am surprised to hear him,

FIGURE 10.22   Electronic Course "The Case Detective".

sitting alone at the computer, answering Andrey Ivanovich's questions, arguing with Vasya". One teacher wrote to us: "My children keep asking me why Vasya, Lisa, and Anfisa speak only with the teacher and with each other?" The author of another letter, a boy, was terribly angry: "Why doesn't Vasya answer me? And Andrey Ivanovich doesn't answer either! And I have so many questions!" As we can see, the children showed willingness to communicate with pedagogical agents, but the limited capabilities of AI did not allow them to fulfill their needs.

The electronic resources described previously were addressed to Russian children studying their native language, but the idea of using a similar resource for studying foreign languages, on the basis of AI, promises to be very fruitful. We consider it very important that such lessons can be used for independent, preparatory work of students with subsequent collective communication activities in the face-to-face mode or online classes, organized on the basis of this preparatory work. Characters, whose lives and personalities the students will get to know during independent work, will also be present in the lesson with the teacher: they will become a topic for stories, discussions and give the opportunity to play their roles, first acting out their dialogues and then creating similar dialogues in similar communication situations. Such resources will undoubtedly facilitate the work of foreign language teachers working in the DL format, as well as time spent preparing for classes and checking homework.

Is it realistic today to use AI to create digital courses for teaching foreign languages with intelligent pedagogical agents who can communicate with learners?

This question can be answered positively. We are all familiar with the programs embedded in our personal devices and docking stations as intelligent voice assistants that can solve various tasks such as to find convenient routes, order a taxi, read bedtime stories, quickly search the Internet for necessary information, refer to articles in dictionaries and encyclopedias and talk with the user on any topic. But creating a similar animated, talking pedagogical agent is not an easy task, not to mention that it is very expensive. It would require the collective work of a whole team of teaching methodologists, artists, programmers and actors to voice its speech. But it turned out to be quite possible as our experience, and the experience of other foreign language teachers, shows the possibility of using ready-made ones, although they were created for other purposes.

Since about 2010, scientific literature, including literature on teaching foreign languages, has been putting forward new pedagogical concepts, technologies, methods and approaches within the pedagogical concept of smart learning. Scientists are researching a wide range of software applications (running on users' smart devices) and considering their benefits (Alanazi, 2019; Al-Mubireek, 2020; Middleton, 2015; Shraim & Crompton, 2020).

We found the work of the American researcher Joshua Underwood (2017), (Arkhangelskaya, A. L., & Dunaeva, L. A., 2007). who taught English to foreign children from 7 to 12 years old, to be the closest to the topic of our research— teaching a language during the initial stage of learning. In his writings, the scientist notes the obvious advantages of using voice assistants in the process of teaching pronunciation, vocabulary and basic speech structures ("What is the weather like today in. . .?", "What time is it now?", "What's your name?", etc.). The experiment carried out by the scientist revealed a high degree of motivation of students in such work, who:

> were intrigued by what any particular implementation of AI was capable of and highly motivated to test their ideas through trial and error, thus engaging in extensive interactive target language speaking whilst developing potentially valuable 21st-century skills for working with AI. Speaking to AIs engaged . . . natural curiosity, largely in order to explore what the AIs were and were not capable of.
>
> (Underwood, 2017, p. 320)

Russian voice assistant Alice caught our attention immediately after it was presented by Yandex in 2017, and we began developing a system of tasks for the tutorial (Rudenko-Morgun et al., 2019) that was designed to work within the BL model, described in the first section of our chapter. The concept of this system of tasks and its experimental verification were covered in detail in scientific articles in Russian and foreign journals (Al-Kaisi et al., 2019, 2021). Here, we want to tell you how we have used the voice assistant (Alice) in the development of the digital course "Russian with Russian friends".

As we have already said, using the Web Speech API speech recognizer to prepare the students for communication in online classes did not completely satisfy us, since it allowed us to train only the memorized replicas of the characters of the digital course. But we wanted to achieve live speech output. What attracted us to Alice was that when she wrote down questions and answers, she did it correctly, not only in terms of grammar and spelling but also with respect to punctuation. Figure 10.23 demonstrates that. Unfortunately, this feature is not available in the Web Speech API. Additionally, a pleasant surprise awaited us. Since we started working with Alice when developing the introductory phonetic course, she acquired a new skill: she learned foreign languages. So, we were able to include a new type of task in the course: ask Alice how the words, phrases and sentences of the lesson are translated into the students' native languages (English, Spanish, Arabic, Chinese, etc.). In Figure 10.24 Alice is requested to translate a word combination from Russian into English.

"Conversations with Alice" have become a good addition to the course. Thematically, they are related to the plots of the lessons. For example, in the first section "Acquaintance", students are invited to: get acquainted with the voice assistant by asking where Alice lives, where she works, if she has a family, friends, a pet and so on; as well as such requests as asking Alice to show photographs of the history faculty of Moscow State University, where Ivan studies, and the Central Children's Library, where Marina works (Ivan and Marina are the characters of the course). This is how students master the vocabulary of the lesson and immerse themselves in the environment where the characters of the digital course live. In the course's second section, shopping, students ask Alice questions about the groceries bought by Ivan, finding out what they look like and what their real price is (prices given in the lessons of the course are purely conditional and educational). Figures 10.25 and 10.26 show dialogues with Alice.

It is important for students to know whether Alice understands them or not. If she does, it means they formulated and pronounced their question correctly. But it

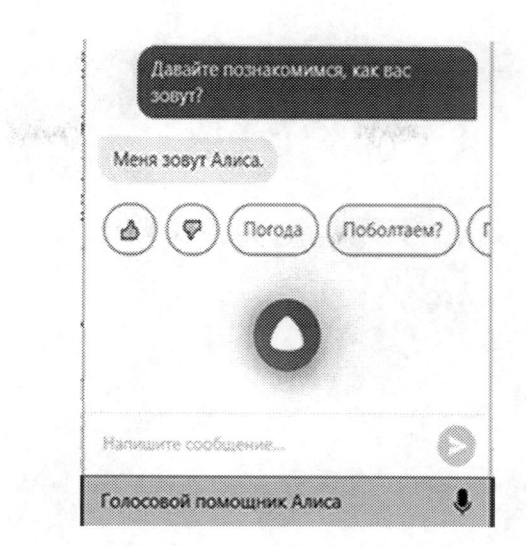

FIGURE 10.23 Pedagogical Agents in the Electronic Speech Course.

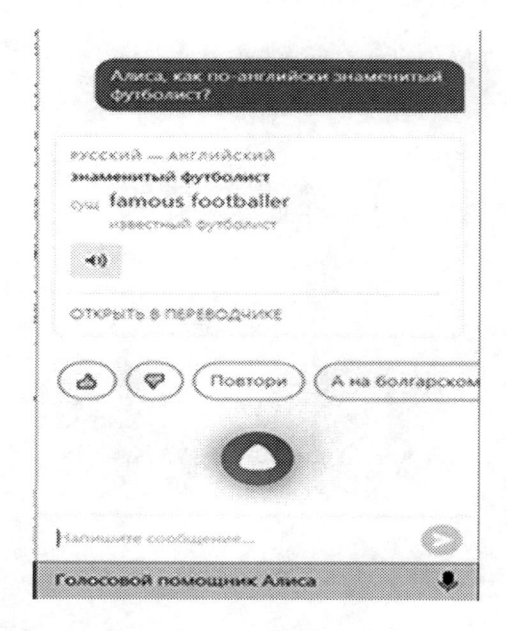

FIGURE 10.24 A Chat with the Voice Assistant Alice. "My name is Alice".

is necessary to warn them that they will not be able to understand everything that Alice says, since she is free in her choice of vocabulary and grammatical structures and she answers the same question differently every time. Students should imagine that they are in Russia and are talking with a native speaker. They may not

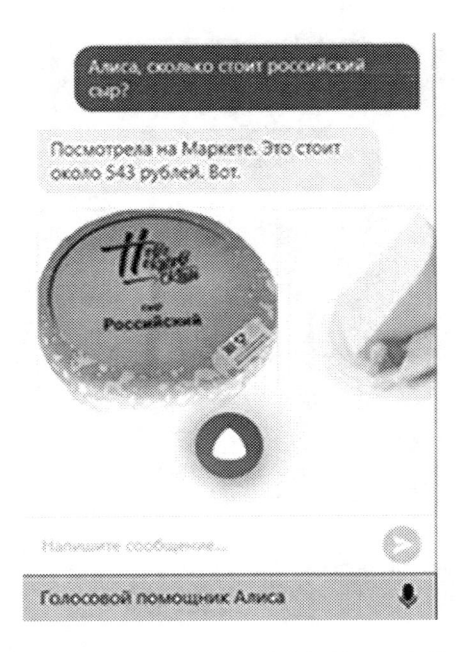

FIGURE 10.25   A Chat with the Voice Assistant Alice. Phrases in Russian and in English.

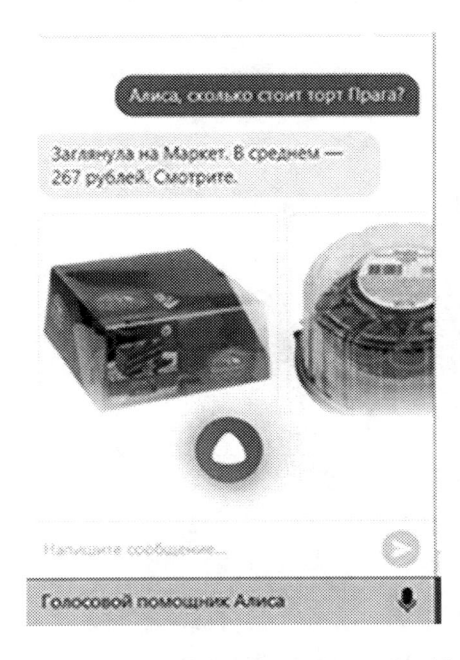

FIGURE 10.26   A Chat in Russian with the Voice Assistant Alice About the Price of Cheese.

be able to understand everything, but they must learn to listen and understand using context—this is how the teacher should instruct the students, inviting them to talk to Alice. It is extremely important to follow this principle when teaching listening and speaking skills. Dealing only with educational texts, students find themselves completely helpless when entering the target language environment; they experience shock and disappointment and lose motivation. Therefore, already at the very early stage of language learning, it is necessary to use authentic material that is carefully measured out. The developer of the course should select the simplest questions that require an accurate and informative answer. In Figure 9.16, we can see that Alice answered the first question simply enough that it can be understood by foreigners who are just starting to learn Russian, and when answering the second question she used vocabulary that was clearly unfamiliar to them: "заглянула" (a colloquial synonym for the verb "looked"), "в среднем" ("on average"). But our goal here was for students to hear and see the numerals, and when discussing a conversation with Alice in an online lesson, to be able to use certain part of speech to tell what question they asked her and what answer they received.

An experiment that we conducted in December 2020, after the students completed the A1 level, gave representative results. Control and experimental groups were formed in accordance with the principles described earlier in section one. Students of experimental group A, when passing the Speaking test, received significantly better marks than those of control group B, which started its training several months earlier, when tasks using the Web Speech API speech recognizer and conversations with Alice were not yet incorporated into the course content. Figure 10.27 shows

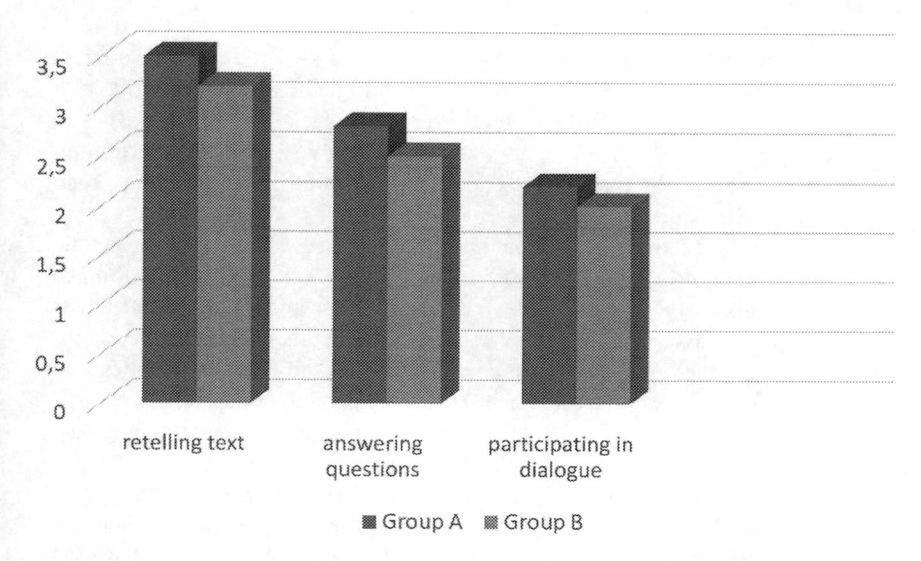

FIGURE 10.27   A Chat in Russian with the Voice Assistant Alice About the Price of Cake.

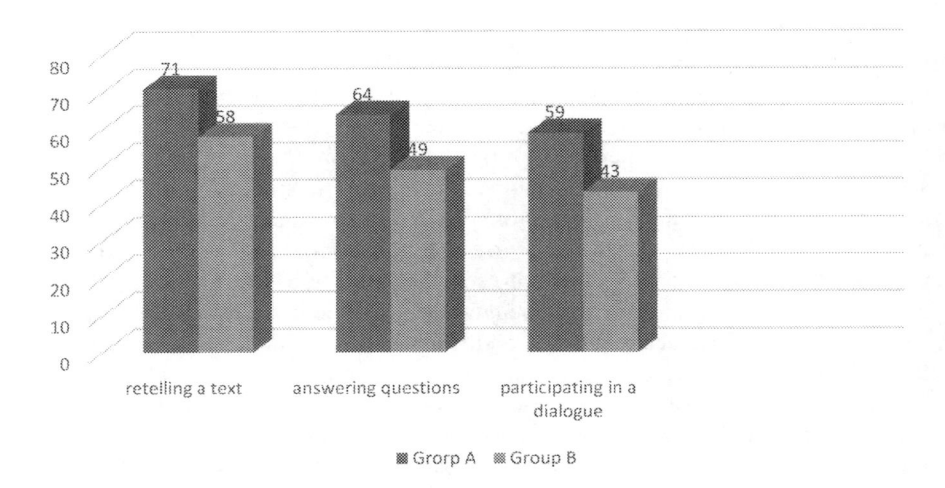

FIGURE 10.28 Results of Speaking Test (Level A1).

the results of this experiment. The digital scale on the left shows the average scores received by group A and group B.

In addition, the teachers who conducted the testing noted the relaxed attitude of group B students, the quick response to questions and the use of speech etiquette formulas in their answers.

In conclusion, it should be said that the study of the problems of using AI for teaching a foreign language has not yet received proper development, but it is obvious that this direction is very promising. Some important conclusions can already be drawn:

- it is necessary to distinguish between the tasks of setting the correct pronunciation and the formation of speaking skills: to solve the first problem, though not fully, speech recognizers such as Google Speech API and Web Speech API can be used; to solve the second problem, attention should be paid to the capabilities of intelligent assistants such as Alice, Siri, and so on;
- at present, speech recognizers and intelligent assistants do not fully meet the requirements of language teaching methodology. Universities cannot make adjustments to the principle of voice recognition and cannot create their own libraries that correspond to the level of their students. Therefore, methodologists developing language courses should study the capabilities and disadvantages of existing tools in order to create, on their basis, correct tasks that meet the didactic goals and objectives of a certain stage of language training;
- AI specialists need to be tasked with the development, in cooperation with methodologists, of tools that meet the needs of teaching a foreign language as they will be widely demanded by Generation Z, who live in the digital world and require new, interactive teaching tools.

REFERENCES

Ahmed, H. D., & Asiksoy, G. (2018). Flipped classroom in language studies: A content analysis of recent articles. *Near East University Online Journal of Education, 1*(1), 11–19.

Alanazi, A. M. (2019). E-assessment through tablets and smart phones: An attitudinal assessment o teachers. *Jeri-International Journal of Educational Research and Innovation, 11*, 1–17.

Al-Kaisi, A., Arkhangelskaya, A., & Rudenko-Morgun, O. (2019, March 11–13). *Voice assistants as a training tool in a foreign language class*. In INTED2019 Proceedings: Paper presented at 13th International Technology, Education and Development Conference, pp. 1236–1246. Valencia, Spain. doi: 10.21125/inted.2019.0399

Al-Kaisi, A., Arkhangelskaya, A., & Rudenko-Morgun, O. (2021). The didactic potential of the voice assistant "alice" for students of a foreign language at a university. *Education and Information Technologies, 26*(1), 715–732. doi: 10.1007/s10639-020-10277-2

Al-Mubireek, S. (2020). Teacher perceptions of the effectiveness of using handheld devices in Saudi EFL classroom practices. *International Journal of Emerging Technologies in Learning, 15*, (22), 204–217. https://doi.org/10.3991/ijet.v15i22.16689.

Archibald, M. M., et al. (2019). Using zoom videoconferencing for qualitative data collection: Perceptions and experiences of researchers and participants. *International Journal of Qualitative Methods, 18*, 1–8.

Baker, J. W. (2000, April 12–15). The "classroom flip": Using web course management tools to become the guide by the side. In *Selected papers from the 11th international conference on college teaching and learning* (pp. 9–17). Jacksonville: Florida Community College.

Banditvilai, C. (2016). Enhancing students' language skills through blended learning. *Electronic Journal of E-Learning, 14*(3), 220–229.

Bergmann, J. S., & Sams, A. (2012). *Flip your classroom: Reach every student in every class every day*. Washington, Eugene, OR—Alexandria, VA: International Society for Technology in Education.

Bhatt, S., Jain, A., & Dev, A. (2020). Acoustic modeling in speech recognition: A systematic review. *International Journal of Advanced Computer Science and Applications, 11*(4), 397–412.

Buddhima, N. W., & Keerthiwansha, S. (2018). Artificial Intelligence education (AIEd) in English as a second language (ESL) classroom in Sri Lanka. *International Journal of Conceptions on Computing and Information Technology, 6*, 2345–9808.

Bulanov, A. (2019, November 11). Speech hour: Speech therapist robot will deal with stroke recovery: New medical device will teach to speak again of victims of cerebral hemorrhage. *Izvestiya* [Chas rechi: robot-logoped zajmetsya vosstanovleniem posle insul'ta: Novoe medicinskoe ustrojstvo nauchit zanovo govorit' postradavshih ot krovoizliyaniya v mozg] (in Russian). Available at: https://iz.ru/946438/aleksandr-bulanov/chas-rechi-robot-logoped-zaimetsia-vosstanovleniem-posle-insulta (Accessed: 23 July 2021).

Chen, J., Lyell, D., Laranjo, L., & Magrabi, F. (2020). Effect of speech recognition on problem solving and recall in consumer digital health tasks: Controlled laboratory experiment. *Journal of Medical Internet Research, 22*(6). https://doi.org/10.2196/14827.

Christopher, J., & Marsh, D. (2016). The flipped classroom. In M. McCarthy (Ed.), *The Cambridge guide to blended learning for language teaching* (pp. 55–67). Cambridge: Cambridge University Press.

de Vries, T. J. (2021, April 14). The pandemic that has forced teachers to go online. Zooming in on tips for online teaching. *Frontiers in Education*. https://doi.org/10.3389/feduc.2021.647445. pdf.

Dodigovic, M. (2007). Artificial Intelligence and second language learning: An efficient approach to error remediation. *Language Awareness, 16*(2), 99–113. doi: 10.2167/la416.0.

Fandej, V. A. (2012). *Theoretical and pragmatic foundations of using the form of blended learning of a foreign (English) language in a language university.* Moscow, PhD diss. [Teoretiko-pragmaticheskie osnovy ispol'zovanija formy smeshannogo obuchenija inostrannomu (anglijskomu) jazyku v jazykovom vuze] (in Russian).

Graham, C. R. (2006). Blended learning systems: Definition, current trends, and future directions. In C. J. Bonk & C. R. Graham (Eds.), *Handbook of blended learning: Global perspectives, local designs* (pp. 3–21). San Francisco, CA: Pfeiffer Publishing.

Iftikhar, G. (2020, March). Coronavirus pandemic reshaping global education system? *AA Anadolu Agency, 19.* Available at: www.aa.com.tr/en/education/coronavirus-pandemic-reshaping-global-education-system/1771350 (Accessed: 21 July 2021).

Johnson, C., & Marsh, D. (2016). The flipped classroom. In M. McCarthy (Ed.), *Blended learning for language teaching* (pp. 55–68). Cambridge: Cambridge University Press.

Johnson, W. L., Rickel, W. J., & Lester, C. J. (2000). Animated pedagogical agents: Face-to-face interaction in interactive learning environments. *International Journal of Artificial Intelligence in Education, 11,* 47–78. http://citeseerx.ist.psu.edu/viewdoc/download?doi=10.1.1.7.7812&rep=rep1&type=pdf. pdf.

Karataş, T., & Tuncer, H. (2020). Sustaining language skills development of pre-service EFL teachers despite the COVID-19 interruption: A case of emergency distance education. *Sustainability (Switzerland), 12*(19), 1–34. https://doi:10.3390/su12198188.

King. (1993). From sage on the stage to guide on the side. *College Teaching, 41*(1), 30–35 https://doi.org/10.1080/87567555.1993.9926781.

Kitishat, A. R., Al Omar, K. H., & Al Momani, M. A. K. (2020). The Covid-19 crisis and distance learning: E-teaching of language between reality and challenges. *Asian ESP Journal, 16,* (51), 316–326.

Leong, L.-M., & Ahmadi, S. M. (2017). An analysis of factors influencing learners' English speaking skill. *International Journal of Research in English Education, 2*(1), 34–41. https://doi.org/10.18869/acadpub.ijree.2.1.34.

Loranc-Paszylk, B., Hilliker, S. M., & Lenkaitis, C. A. (2021). Virtual exchanges in language teacher education: Facilitating reflection on teaching practice through the use of video. *TESOL Journal, 12*(2). https://doi:10.1002/tesj.580.

Lutskovskaia, L., Shoustikova, T., & Udina, N. (2019). AI-based tools for foreign language training: Opinions from different audiences. *Journal of Critical Reviews, 6*(6), 404–409. https://doi.org/10.31838/jcr.06.06.61

Marshall, D., & Ward, L. (2020). Let's collaborate! technology, literacy, and teaching during a pandemic. *Technology and Engineering Teacher, 80*(1), 30–31.

Maureen, J., Lage, G. P., & Michael, T. (2000). Inverting the classroom: A gateway to creating an inclusive learning environment. *The Journal of Economic Education, 31*(1), 30–43.

Mazur. (1997). *Peer instruction: A user's manual.* Harlow: Pearson Education Limited and Prentice Hall.

Medvedeva, M. S. (2015). *Formation of the readiness of future teachers to work in blended learning.* Moscow, PhD diss. [Formirovanie gotovnosti budushhih uchitelej k rabote v uslovijah smeshannogo obuchenija] (in Russian).

Metatla, O., et al. (2019, May). Voice user interfaces in schools: Co-designing for Inclusion with visually-impaired and sighted pupils. Conference on human factors in computing systems—Proceedings. *Association for Computing Machinery, 378,* 1–15. https://doi.org/10.1145/3290605.3300608.

Middleton, A. (2015). *Smart learning: Teaching and learning with smartphones and tablets.* Sheffield: Sheffield Hallam University.

Mohammadi, G., Pezeshki, F., Mohammadhosseinzadeh Vatanchi, Y., et al. (2021). Application of technology in educating nursing students during COVID-19: A systematic review. *Frontiers in Health Informatics, 10*(1), 64.

Oakes, E. H. (2007). *Encyclopedia of world scientists*. New York: Infobase Publishing.

Passov, E. I. (1991). *Communicative method of teaching foreign language communication*. Moscow: Prosveshchenie [Kommunikativnyj metod obucheniya inoyazychnomu obshcheniyu] (in Russian).

Picciano, A. G. (2009). Blending with purpose: The multimodal model. *Journal of Asynchronous Learning Network*, *13*(1), 7–18. https://doi.org/10.24059/olj.v13i1.1673.

Ross, B., & Gage, K. (2006). Global perspectives on blended learning: Insight from WebCT and our customers in higher education. In C. J. Bonk & C. R. Graham (Eds.), *The handbook of blended learning: Global perspectives, local designs* (pp. 155–168). San Francisco, CA: Pfeiffer Publishing.

Rudenko-Morgun, O. I., & Arkhangelskaya, A. L. (2006). *1C: School. Morphemics. Word formation 5–6 grades. Electronic tutorial for schools*. Moscow: 1C Publishing. CD-ROM [1C: Shkola. Morfemica. Slovoobrazovanie. 5–6 klassy. Elektronnoe uchebnoe posobie dlya shkol.] (in Russian).

Rudenko-Morgun, O. I., Arkhangelskaya, A. L., & Al-Kaisi, A. N. (2015). *1C: School. Lexicology. 5–6 grades. Electronic tutorial for schools*. Moscow: 1C Publishing. CD-ROM [1C: Shkola. Leksikologiya. 5–6 klassy. Elektronnoe uchebnoe posobie dlya shkol.] (in Russian).

Rudenko-Morgun O. I., & Arkhangelskaya, A. L. (2016). *1C: School. Morphemics. Word formation. 5-6 grades. Electronic tutorial for schools*. Moscow: 1C Publishing. [1C: Shkola. Morfemica. Slovoobrazovanie. 5 – 6 klassy. Elektronnoe uchebnoe posobie dlya shkol.] (in Russian).

Rudenko-Morgun, O. I., Arkhangelskaya, A. L., & Al-Kaisi, A. N. (2017). *Studding Russian ourselves and at the lesson. Russian cases: Russian as a Foreign language tutoral book: Level A2*. Moscow: RUDN. [Uchim russkij samostoyatel'no i na uroke. Ruskiye padezi. Uchebnoe posobie: Uroven' A2]. (in Russian).

Rudenko-Morgun, O. I., Arkhangelskaya, A. L., & Al-Kaisi, A. N. (2019). *Studding Russian ourselves and at the lesson. Introductoty phonetics and grammar course tutoral book: Level A1*. Moscow: RUDN. [Uchim russkij samostoyatel'no i na uroke. Vvodnyj fonetiko-grammaticheskij kurs. Uchebnoe posobie: Uroven' A1] (in Russian).

Rudenko-Morgun, O. I., Arkhangelskaya, A. L., & Dunaeva, L. A. (2007). *Russian language with computer. Step 1: Computer course*. Moscow: RUDN—Istrasoft—1C. CD-ROM.

Salieva, Z. (2020). The challenges of distance learning in the period of pandemic: The case of teaching speaking. *International Journal of Advanced Science and Technology*, *29*(7), 2112–2116.

Saxena, S. (2013, November 2). How to best use the class time when flipping your classroom? *EdTeachReview*. Available at: https://edtechreview.in/trends-insights/insights/726-how-to-best-use-the-class-time-when-flipping-your-classroom (Accessed: 21 July 2021).

Shraim, K., & Crompton, H. (2020). Perceptions of using smart mobile devices in higher education teaching: A case study from palestine. *Contemporary Educational Technology*, *6*(4). https://doi.org/10.30935/cedtech/6156.

Stein, J., & Graham, C. R. (2020). *Essentials for blended learning: A standards-based guide*. New York: Routledge. https://doi.org/10.4324/9781351043991.

Suhov, K. (2013). Web speech API—miracle from HTML5. Speech recognition on a web page. *System Administrator*, *1–2*, 108–111 [Web Speech API—chudo ot HTML5 Raspoznavanie rechi na veb-stranice. *Sistemnyj administrator*] (in Russian).

Tam, G., & El-Azar, D. (2020). 3 Ways the coronovirus pandemic could reshape education. *World Economic Forum*, *1–5*. www.weforum.org/agenda/2020/03/3-ways-coronavirus-is-reshaping-education-and-what-changes-might-be-here-to-stay/ (Accessed: 21 July 2021).

Teoh, G. B. S., Lin, A. L. W., & Belaja, K. (2016). Which aspects of the english language do distance learners find difficult? *Turkish Online Journal of Distance Education*, *17*(3), 111–119. https://doi.org/10.17718/tojde.19593.

Thorne, K. (2003). How to support blended learning. In K. Thorne (Ed.), *Blended learning: How to integrate online & traditional learning* (pp. 19–24). London: Kogan Page.

Tregubov, V. N. (2020). Using voice assistants to develop English scientific speech. *International Journal of Open Information Technologies, 8*(6), 62–71 [Ispol'zovanie golosovyh assistentov dlja razvitija anglijskoj nauchnoj rechi]. (in Russian).

Tseng, H., & Walsh, E. J. (2016, April). Blended vs. Traditional course delivery: Comparing students' motivation, learning outcomes, and preferences. Quarterly review of distance education. *Quarterly Review of Distance Education, 17*, 43–52.

Tucker, C. R., Wycoff, T., & Green, J. T. (2016). *Blended learning in action: A practical guide toward sustainable change.* Thousand Oaks, CA: Corwin.

Turchi, L. B., Bondar, N. A., & Aguilar, L. L. (2020). What really changed? Environments, instruction, and 21st century tools in emergency online English language arts teaching in United States schools during the first pandemic response. *Frontiers in Education, 5.* https://doi.org/10.3389/feduc.2020.583963.

Underwood, J. (2017). Exploring AI language assistants with primary EFL students. In L. B. K. Borthwick (Ed.), *CALL in a climate of change: Adapting to turbulent global conditions—short papers from EUROCALL* (pp. 317–321). doi: 10.14705/rpnet.2017. eurocall2017.733.

Vasilyeva, T. V., Vlasov, E. A., & Rudenko-Morgun, O. I. (1991). *Case detective.* Moscow: Langsoft LTD, Inbound. CD-ROM [Padezhnyj detektiv] (in Russian).

Vurdien, R. (2019). Videoconferencing: Developing students' communicative competence. *Journal of Foreign Language Education and Technology, 4*(2), 269–298.

Zawacki-Richter, O., et al. (2019). Systematic review of research on Artificial Intelligence applications in higher education—where are the educators? *International Journal of Educational Technology in Higher Education, 16*(39), 1–27. https://doi.org/10.1186/s41239-019-0171-0

11 Impact of Artificial Intelligence (AI) and Robotics on Students' Career Choice

Muhammad Mujtaba Asad, Salma Idrees, Fahad Sherwani, Zafarullah Sahito and Al-Karim Datoo

CONTENTS

11.1 INTRODUCTION

The world is growing incredibly fast and people need to update themselves every single day with the newest technologies and latest innovations. Not only is there a need to acquire new technological products; one must also update oneself on the know-how to use these technologies in daily life without hesitation or fear. Additionally, recent innovations and machines are adding new comforts in people's lives and helping people to become more knowleadgeable. In a nutshell, the world is becoming competent enough to harness technology for the benefits of human beings and society overall. Information Technology (IT) started its journey from very simple machines and has grown in the fields of Artificial Intelligence (AI), Robotics, Databases, Human

Computer Interactions (HCI) and many more. Human beings have now reached the level of creating intelligent machines with comparable intelligence to human beings (Shahzad, 2018).

AI would not have been invented if not for the biologists' discovery that neurons in the human brain fire signals (Teng, 2019). This discovery inspired the idea that computers can also simulate brain functions like human brains. Consequently, the idea aspires towards something that can be as smart as the human mind. The idea is quite simple: whenever a neuron receives a signal, even if it fires the signal back or not, it connects the understanding to a binary system. The same concept was successfully incorporated and employed in computers too. However, the phenomenon in brains is different than in machines, because when the human brain fires signals, neurons get more closely linked mechanically rather electronically (Teng, 2019). With much work and gradual progress, scientists came up with inventions such as Artificial Intelligence (AI), Robotics, Databases and Human Computer Interactions (HCI). Aside from all the details, advantages, disadvantages and its history, one thing is very important and unique about AI—when comparing both natural intelligence and Artificial Intelligence, especially in highly repetitive tasks, AI beats human beings by its efficiency and accuracy. It can also outcompete in complex tasks such as image detection and video analysis. To sum up, it would be an understatement to say that AI is playing a key role in today's society.

11.2 WHAT IS ARTIFICIAL INTELLIGENCE (AI) AND ROBOTICS?

Artificial Intelligence (AI) is a controversial but very interesting field of Robotics. The working of the robot is systematic and it is designed to be augmented from the point of possibilities (Chaus, 2019). The term "intelligence" refers to mind's general capability to solve problems, reason out consequences and understand situations. It is the combination of many skills, knowledge and various cognitive functions like calculations, language, perception and memory (Shabbir & Anwer, 2018). The term "artificial" refers to something that is a manmade fascimile of a real thing; something created by human beings.

Additionally, Artificial Intelligence (AI) in general terms refers to intelligent computer systems that have human designed intelligence, that can perceive things and can perform many cognitive functions that were (in the past) only associated with the human mind. In short, AI has the ability to simulate processes of the human mind and human intelligence. Wisskirchen (2017) in his book *Artificial Intelligence and Robotics and Their Impact on the Workplace* divided AI in two categories: Weak Artificial Intelligence (WAI) and Strong Artificial Intelligence (SAI). WAI refers to those computers that only inquire cognitive processes and replicate intelligence, whereas SAI refers to those computers which also have self-learning processes. Those which have ability to understand their functions and can improve their abilities by reflecting its past experiences. WAI has the ability to focus on one specific task; in contrast, SAI has the ability to focus on more than one specific area (Siau & Yang, 2017). The dictionary definition of Robotics is the branch of technology, engineering and science that deals with design, creation, working, programming, application and practicality of robots. Robots may look like machines or humans depending upon function and application ("Dictionary", n.d.). In addition, robots perform various tasks without human control or assistance.

11.3 A BRIEF HISTORY OF ARTIFICIAL INTELLIGENCE (AI)

The beginning of AI was based only on philosophy, fiction and imagination. Later, inventions like electronic machines, calculating machines, and computers influenced the idea of AI. In early years, mastery in playing chess was the most intelligent task an I could perform. In 1789, Freiherr Joseph Friedrich zu Racknitz created a chess-playing machine named *"The Turk"* which was then considered as intelligent machine. The machine was designed in a way that it could play chess with chess player autonomously. However, it was not a true artificially intelligent, autonomous machine because there was a big game behind the machine. It was actually being operated by a human chess master who was hidden within the machine. Though the machine was not autonomous, the idea behind that machine was amazing and clever; it later contributed to the studies of AI. In the early 1900s, a book named *Tik-Tok of Oz* written by L. Frank Baum also inspired many AI researchers to create something like the character of the book who was an "extra-responsive, thought-provoking mechanical man with the ability to think, speak, act and do everything but live".

Studies began, and in the 1950s, a paper was presented which was a major turning point in the field of AI (McCarthy et al., 2006). The paper was about an electronic computer with such programming as to make it behave intelligently. Then in 1963 a book by Feigenbaum and Feldman (1963) containing collections and descriptions of working of AI programs was published (Buchanan, 2005). Recently, AI technology has aided in the design of communication technologies such as satellites, which can orbit with its 486 processor in the space. Inventors claim that self-replication of AI could easily be created outside the world, and human will not get any chance to fight in vacant spaces on equal terms (Shabbir & Anwer, 2018). These contributions all helped provide ways for AI researchers to investigate more and more about AI and robotics.

11.4 FOUR INDUSTRIAL REVOLUTIONS (IR)

The trend that began in 18th-century Western Europe and has since prevailed worldwide can be considered now the fourth Industrial Revolution (IR). In the book (Wisskirchen, n.d) authors briefly described all four IRs. The first one was "Industry 1.0: Industrialization" when for the very first time, steam engines were introduced. Those steam engines enabled machines to produce goods and services. The second IR was "Industry 2.0: Electrification"; the era when machines became ubiquitous in factories and automatic productions. For the first time, automatically manufactured goods and services were transported to different countries and continents. The third IR was "Industry 3.0: Digitalization", indicating the era of Information Technology (IT) and further electronic automations. In this era, the internet was at its initial stage and use of it made peoples' lives easier, machines replaced human labors and global access to information became easier during this period. The fourth and recent IR is "Industry

4.0", where the world has reached to the level of cyber-physical systems, AI (cognitive computing), robotics, Human Computer Interactions (HCI), the Internet of Things, cloud storages, autonomous vehicles, 3-D printing, nanotechnology, biotechnology, materials science, energy storage, quantum computing and creating smart factories. In a nutshell, the first IR was the period of water and steam power engines, the second IR was the period of electric power to gear mass production, the third IR was the period of electronics and information technology and the latest fourth IR is built based on the third IR, a period of fusion of technologies that is overlapping physical, digital and biological domains. As now we are living in the age of fourth IR, Industry 4.0, the world faces many challenges in the future. At the same time, because of the fourth IR, there are a lot of opportunities to avail to keep oneself up to date with the world.

11.4.1 INDUSTRY 4.0

Industry 4.0 has been given different names by different countries. This name Industry 4.0 was given by Germany (Howaldt et al., 2017). The main concept of Industry 4.0 is "The Industrial Internet of Things (IIoT)", and it talks about the integration of the Internet of Things (IOT), and technologies with industrial production, resulting in digitized links to the creation of industrial value (Kiel et al., 2017). The same concept of Industry 4.0 is being used worldwide with different names as highlighted before. The United States used a similar concept—"Internet Consortium in USA" (Müller et al., 2017), "The internet plus or Made in China 2025" (Keqiang, 2016) and "Manufacturing Innovation 3.0 in South Korea" (Kang et al., 2016). This concept represents systems that work in real-time and are capable, intelligent, horizontal, with vertical and 360 degrees of connection of people, and in communication with other dynamic and complex systems (Abramovici & Herzog, 2016). The Industry 4.0 or IOT has many characteristics like Cyber-Physical Systems that are comprised of sensors and data processing units that represent the fusion of the real and virtual worlds and enable real time data transfer. Human to human, human to object, and object to object interactions are made possible by Industry 4.0. The aim of the IoT and Industry 4.0 is also to address contemporary issues (e.g., shortened technology and innovation cycles, increasing customization in technology, and enhanced demand volatility) the world is facing and will face in the future, based on these characteristics and many others.

11.5   INTEGRATION OF AI AND ROBOTICS IN EDUCATION (IN GENERAL), IN COMPUTER SCIENCE AND ENGINEERING (IN PARTICULAR)

As in many fields, AI and Robotics have influenced the field of education through different means. During the late 1990s, Robots first appeared in US classrooms for educational purposes (Bers & Portsmore, 2005). Virtual Pedagogical Agents presented in mobile phones, tablets, laptops and other gadgets and Intelligent tutoring systems (ITSs) were also invented to deliver education with valid comprehensive reviews. Social Robots have also been recently explored for the field of education for benefiting the iteaching and learning process. Hence, AI and robotics have now taken the places of teachers and assistants.

Moreover, two to three years back a survey was conducted on the topic of "Long Term Human Robot Interaction (HRI)". The main reason behind this survey was the popularity of using Robots and intelligent machines. The objective was to see the usage of AI not generally but in real classrooms. The motive was to invent and integrate "Social Robots" in the field of education. Soon after this occurred many other restricted surveys conducted in the same domain (Belpaeme et al., 2018).

It is very important to keep in mind the social aspects of the learning experience through interaction among teachers and learners. Artificially intelligent robots can teach, but still the real pedagogy can not be expected from robots. It is very true that robots can be used as pedagogical tools for Science, Technology, Engineering and Mathematics (STEM) education, but robots themselves cannot think of pedagogies and strategies based on the situations and needs of individual learners.

Robots in the education field can be a controversial topic. For example, how can a robot achieve the learning outcomes set for the teaching-learning process? How can these machines overcome critical situations that may appear? How can a robot handle students' emotional well-being, psychological needs and diversity? In spite of these questionable scenarios, if robots are really being able to integrate into educational sectors as a tutors or teachers, we as a society can get number of benefits overall. As a result, robots will then become one of the prime resources of providing direct curriculum support via hints, tutorials and supervisions.

Apart from social robots, the term "virtual agents" is also used by educators (Belpaeme et al., 2018). A virtual agent is basically a computer-generated, artificially-made virtual character that helps as an online teacher. Its functions include intelligent conversations with users, responding to queries and also performing non-verbal behaviours when needed for the users. Though there are many benefits, as discussed, direct physical interaction is not possible with these virtual agents. In contrast, "Social Robots" can interact directly with learners, as their use is now common in some parts of the world. Robots with visual agents can be more engaging, leave positive impact on learning and be more enjoyable. Studies (Belpaeme et al., 2018) highlighted a proposal that robots aid individuals with visual impairment for children under the age of two and four, with minimal learning. These robots will provide educational trainings to these kids via screens. Similarly, Balpaeme described how in the early years, teaching assistant robots were invented that were made to observe classrooms with the aim of improving and having a qualitative impact on learners' attitudes and progress. Additionally, a commercial tutor robot named "IROBI (Yujin Robotics)" was released in the early 2000s. This robot was designed to teach English. IROBI was able to enhance both students' concentration levels during learning activities and their academic performances, as compared with other technologies like virtual agents. There are also many other inventions that have made education easy to acquire. Some robots were invented for younger learners in scientific areas, like language learning and developmental psychology. Additionally, a fully autonomous robot was developed for nursery schools to develop vocabulary skills of 18-to-24-month-old toddlers. These robots functioned in a way that they themselves were involved in one-to-one interactions for personalized education. In some cases, the robots were used as a path through which a lecture is delivered; they were not interacting in a way a teacher interacts, but they were acting like an assistant for the teacher. In the technological world, robots not only work as tutors or teachers but also as what is known as peer robots or learning company robots. "Robovie" was the first ever fully autonomous robot that was introduced in elementary school. It was an English-speaking peer robot for Japanese students. It was invented to improve students' English language. The robot was used for a few weeks and later the results were disclosed, with the conclusion that students got accurate responses from these peer robots as compared to identical looking tutor robots (Belpaeme et al., 2018).

Nevertheless, in the early 2000s, after-school autonomous robotics programmes and robotics competitions named "ROBOFEST" were launched to combine computers and robots. In the exhibitions, students promoted more creativity than fixed game competition. They invented autonomous robotics that were helpful in formal and informal learning environments used to improve Math and Science learning, adding critical thinking and problem solving skills as well (Matson et al., 2004). This "ROBOFEST" competition, which continues annually to this day, promoted development in robotics-based pedagogy to teach STEM skills. The characteristics of robotics-based pedagogy were directly or indirectly providing five key advantages compared to traditional pedagogy in teaching: (1) integration of STEM topics in multifaceted shapes, (2) effective conversion of abstract concepts into real-life practical learning modules for learners, (3) aligning STEM theory with practice, (4) hands-on active engagement, and (5) enjoyable and inspiring learning environment (Chung et al., 2014). This annual competition challenges teams to create an autonomous robot that supports and naturally associates with STEM education and components. Furthermore, it has now become an international competition engaging teams from 13 US states and other teams from eight countries including India and China. Most of the ROBOFEST games are designed in a way that students learn Math and Science through hands-on robotics educational experiences, with direct links to concepts in Physics and Math.

11.6 IMPORTANCE OF AI AND ROBOTICS IN TODAY'S WORLD REGARDING THE NEEDS OF INDUSTRIAL DEMANDS

The Industrial Revolution 4.0 has reached levels of cyber-physical systems, cognitive computing, Human Computer interaction (HCI), Human Robot Interaction (HRI), the world of cloud storages, Internet of Things and creation of smart factories. AI, robotics and other up-to-date technology are ruling today's world by replacing human beings, market jobs and even teachers with recent robotics and AI machines. Anderson and Smith (2014) in their report highlighted the importance and impact of AI and robotics in today's world, projecting that AI and robotics will take charge of most of daily life by 2025. This will impact a wide range of industries such as transport, logistics, customer care services, health care and even home maintenance. Additionally, AI and robotics will challenge the economic and employment landscape over a next decade.

Today's world is very much influenced by AI; it has become an integral part of our daily lives. We use smart gadgets to navigate around town and virtual digital assistants like Siri, Alexa and Cortana for our queries. As mentioned previously, each and every human institution is figuring out new and innovative ways to use AI to benefit themselves and others, including financial institutions, legal institutions, media companies and insurance companies.

People want an easy life and AI is making it so, such as in helping humans to multitask where it was not possible before. For instance, people can now turn on their home air conditioning systems while driving before reaching home, and they can view the person at their doorbell via a camera while doing their work in their bedrooms. AI is now revolutionizing companies to compete and grow across the world by presenting new production factors that drive business towards record profits. Realizing the importance of AI, business sectors are grabbing up opportunities and developing active AI strategies. Although businesspeople are primarily concerned with whether their products have a market, they also consider their development with moral and ethical questions in mind, which resultantly leads to positive feedback, empowering people with superior innovations.

AI is not just helping us to improve our lives but in some cases it is helping us to save our lives, as AI serves the healthcare systems by providing powerful diagnostic tools for health. With the help of AI, many robots have been invented to assist in different surgeries. AI also addresses today's contemporary challenges and issues, such as global warming. Invention of smart infrastructures, smart appliances and smart agricultural techniques are helping more efficiently with natural resources. AI also provides benefits in fields like engineering, economics, linguistics, law, manufacturing and many more.

11.7   HOW ARTIFICIAL INTELLIGENCE (AI) WILL INFLUENCE THE WORLD IN THE FUTURE

AI seems a hot topic to discuss nowadays because of its ubiquity and impact in every field of life, and this discussion leads to the prevalence of AI in the future. It is changing the world with each single day with new autonomous inventions. AI and Robotics can impact people's lives both negatively as well as positively, especially regarding careers and jobs. Moreover, world-renowned personalities like Stephen Hawking and Bill Gates have predicted mass unemployment due to the rise of Smart Technology, AI, Robotics and algorithms—termed as STARA (Bort, 2014). It was estimated that approximately one-third of the present jobs can be replaced by STARA by 2025 (Frey & Osborne, 2013), all because of updated technology, development in robotics, intelligence, and inexpensive autonomous units that have potential to beat human beings at many conceptual and manual tasks (Frey & Osborne, 2013).

A study on 702 occupations was conducted, and for most of them it was concluded that STARA would replace their occupations. Jobs at risk include accountants, market research analysts, commercial pilots, customer service representatives, sales staff, office/administration workers and more (Frey & Osborne, 2013). STARA will highly affect the fields of medicine (Bloss, 2011), education (replacing teachers with online tutors, virtual agents, robots as teachers, robots as peers and robots as assistant teachers), transportation, farming, forestry and fishing. Websites will be updated automatically webpages will be reformatted automatically based on eye-tracking data (Siau & Yang, 2017). Over all, this study concluded that 47% of jobs worldwide are at risk because of the rise of STARA (Frey & Osborne, 2013). Even jobs that are not fully affected or not at high risk will still be disturbed by STARA. For example, driving instructors and license testing officers may become redundant with driverless cars and insurance assessors (Brougham & Haar, 2018). In addition, researchers also have claimed that vehicles will drive automatically at night so that drivers can sleep well without fear of accidents. Like other forms of technology, AI and robotics also need dedicated training courses to prepare professionals for their utilization. As the number of AI and robotics devices increases, the need for professionals to support and maintain their functioning also increases. Expertise for every invention will be needed to maintain their functioning. Resultantly, there will be a huge need of these kinds of professional in an ever-growing market. As per industry estimates, there was a labor shortage in the field of technology, AI and robotics by 2017. For open developer positions, the United States needed almost 500,000, yet there were only 50,000 computer science graduates to satisfy the demand. In conclusion, keeping in view these figures and focusing on the seriousness of this issue, in the current era of business development dominated by AI and robotics, there will be more demand than ever for tech professionals to rule the market (Chaus, 2019).

11.8 ROLE OF ARTIFICIAL INTELLIGENCE AND ROBOTICS IN CAREER SELECTION

In today's world everyone is busy doing hard work to build a bright career where they can create and maintain a luxurious life for themselves and their loved ones. Career selection lays the foundation of a person's life and how they build their future.

Though a student may set a particular goal in his or her life, that goal may not always remain constant; it changes with the needs of society, the job market and latest technologies and innovations. Some students opt for one field of study but later change their career to meet the needs of job market. Others continue with the same thing they studied. As a high school student, choosing an appropriate career can be a very stressful decision. Similarly, changing career as an adult can also be a difficult transition (Gray et al., 1990). Based on my personal experience with Pakistani students, selecting a career can be a challenging and depressing thing to do. Students often come to know about their passions and skills during their graduation period. Unfortunately, most of them continue with the same career path in which they began, due to the time they invested in it. They can't waste time switching career paths, as this is often considered a bad decision. Students are sometimes unaware of innovative technological developments because their main objective of getting degrees is to earn money, not to keep oneself up to date on these matters.

11.9 THEORIES FOR MOTIVATION FOR CAREER SELECTION

Human being cannot survive in society without work or income. Many people work from a very early age because of their economic conditions, with the motivation of earning money to live a better life. In contrast, some people work to enhance their competencies and skills, choosing the optimal occupation for themselves to secure their careers and future lives. Some work longer into late adulthood than is healthy, due to uncertain economic conditions (Kooij & Kanfer, 2019). Resultingly, a steady increase in age diversity among workforces was experienced by organizations with new challenges. Researchers have designed many theories, studies and models to examine the motivation and determination of individuals towards their careers in the context of larger notions of society. Some of the theories and studies are discussed to follow:

- **Person-Centric Perspective:** This perspective focuses on multiple motivational dynamics of present and future needs. It influences an individual's perception, comprehension and interpretation of the world around them and also influences their role as a worker. The person-centric perspective with its importance on affect, behaviour and cognition (Bandura, 1977), offers possibilities for fruitful incorporation with approaches that focus on work motivation throughout life (Kooij & Kanfer, 2019).
- **Self-Determination Theory:** This theory says self-determination in setting goals and objectives causes highers level of motivation that help an individual to shape them internally. It highlights beneficial consequences of work goals that give birth to a sense of control and achievement by self-efficacy. In a nutshell, this theory helps a person to set a goal around which their core values revolve, which makes a person autonomous and competitive. There are five components of this theory: handling social cognitive career effectively, influencing career-decisions, determining student's career aspirations, developing career choices and setting life goals both intrinsically and extrinsically (Chantara et al., 2011).
- **Self-Concordance Theory:** This theory was based on self-determination theory. It advocates that people choose and select personal goals that match their personal interests. In contrast, this theory further explains that when people set abstract and illogical goals for themselves, they do not do a good job (Sheldon et al., 2020).
- **Job Characteristic Theory:** This theory reflects the idea that there are certain job characteristics; skill variety, task identity and autonomy. These characteristics influence work motivation and affect three critical psychological areas: 1) meaningfulness of work, 2) responsibility for outcomes and 3) knowledge of results. This theory highlighted the socio-relational features of work, because these features impact work motivation and performance (Grant & Parker, 2009).
- **Goal-Setting Model:** The goal-setting model unites goal choice and goal striving. In this theory, goals change from motivational states into action states when facilitated by attention, coordinated effort, increasing task persistence and strategies for goal accomplishment (Beier et al., 2019).
- **Social Cognitive Career Theory (SCCT):** There are two models in SCCT. The new model revolves around seven sets of variables: work satisfaction, affective and personality traits, goal directed activity, overall life satisfaction, self-efficacy, work conditions and outcomes and finally environmental supports and obstacles. In this model, outcome expectations are absorbed between working conditions and outcomes. In a nutshell, it can be summarized as what one wants and what one receives from work. There is also a previous SCCT model from which many of the features were adapted in the new SCCT model like self-efficacy beliefs, outcomes expectations, goals, other personality traits and contextual affordances. The difference between

the two models regard predictive criteria and content (Lent & Brown, 2008). The previous model represents a conceptual comprehensive building for conceptualizing interest formation, career choice, and academic and career performance (Schaub & Tokar, 2005). There was a huge mission in the United States regarding the number of students entering into computing disciplines like computer science, computer engineering, software engineering, information systems and information technology. They wanted to attract students to continue in such fields, along with Science, Technology, Engineering and Mathematics (STEM) fields. SCCT guided research on this topic focusing on career related self-efficacy of the individual. SCCT focuses on the interaction between diverse people, environmental and behavioural variables that impact the process by which individuals (a) enhance academic and career interests, (b) make and revise their educational and skills goals, (c) secure best performance and quality in their academic and career goals (Lent et al., 2008).

11.10   STUDENTS DON'T KNOW WHAT TO SELECT AND WHY TO SELECT?

Researchers have worked on various theories for students to select optimal careers and to motivate them to become self-determined towards their educational and vocational goals. Still, as per my personal experience, students are often unsure of what they really in terms of their careers. They don't know how to set a goal for their favourite particular career, and if they succeed in setting a goal, still they are confused about the reasons for setting that goal. They often don't know that why they are selecting that particular career. Is it really their passion, or just chance that they came to know about a particular career path and chose it? And when it comes to the latest innovations and STEM knowledge, many Pakistani students are unfortunately outdated, not bothering to update themselves with global knowledge and inventions. Students are most of the time unaware of trending jobs, market values, what need to be updated, what skills are required in a particular time period, and most importantly why it is important to consider these factors when selecting a career.

Some theories are discussed in this article and many other theories can be found to get a sense of career choice based on an individual's motivation, interests, their work satisfaction, self-efficacy, their working conditions and many more factors. Students can refer to these theories to find a way in their careers.

11.11   CONCLUSION

Selecting a profession and university is a difficult task for students; perhaps the singlemost important decision of a student's life. This stage must be carefully governed since it shapes the professions of students in the future (Sevindir & Yazici, 2014). In this Industry 4.0, among the spreading domains of technology, many entrepreneurial students aim for careers related to technology. Therefore, students who choose to be computer scientists and engineers get more benefit from this digital age. They are completely aware of the fact that their careers are extremely enmeshed in AI, due to the content they study like abstract algebra, real analysis, complex analysis, topology, intro to C++, C# or perhaps Python, theoretical statistics and numerical analysis. Engineers learn to create machines; their working, functioning and formulation.

In addition, there are various skills and courseworks that train students to work with AI. In Math, these topics include: Statistics, Predictions, Calculus, Algebra, Bayesian, Algorithms and Logics. In Science, they include: Physics, mechanics, cognitive learning theory, language processing. And in Computer Sciences, these topics include: data structures, programming, logic and efficiency. Developing a love for smart technology and machines are key to becoming a skillful AI professional. This chapter indicates that AI systems, robotics and Industry 4.0 are very important innovations in this era of advancement, growth, development, bright careers and alterations. This highlights the importance of the latest technologies in careers, job market and self-advancement. These technologies benefit human beings by making their lives easier and smarter.

REFERENCES

Abramovici, M., & Herzog, O. (2016). *Engineering im Umfeld von Industrie 4.0: Einschätzungen und Handlungsbedarf.* München: Herbert Utz Verlag.

Anderson, J., & Smith, A. (2014). *AI, robotics, and the future of jobs.* Washington, DC: Pew Research Center Report.

Bandura, A. (1977). Self-efficacy: Toward a unifying theory of behavioral change. *Psychological Review, 84*(2), 191.

Beier, M. E., Bradford, B. C., Torres, W. J., Shaw, A., & Kim, M. H. (2019). Cognition, motivation, and lifespan development. In *Work across the lifespan* (pp. 157–177). Amsterdam: Elsevier.

Belpaeme, T., Kennedy, J., Ramachandran, A., Scassellati, B., & Tanaka, F. (2018). Social robots for education: A review. *Science Robotics, 3*(21), eaat5954.

Bers, M. U., & Portsmore, M. (2005). Teaching partnerships: Early childhood and engineering students teaching math and science through robotics. *Journal of Science Education and Technology, 14*(1), 59–73.

Bloss, R. (2011). Mobile hospital robots cure numerous logistic needs. *Industrial Robot: An International Journal.*

Bort, J. (2014). Bill Gates: People don't realise how many jobs will soon be replaced by software bots. *Business Insider* (Accessed: 10 December 2014).

Brougham, D., & Haar, J. (2018). Smart technology, Artificial Intelligence, robotics, and algorithms (STARA): Employees' perceptions of our future workplace. *Journal of Management & Organization, 24*(2), 239–257.

Buchanan, B. G. (2005). A (very) brief history of Artificial Intelligence. *Ai Magazine, 26*(4), 53.

Chantara, S., Kaewkuekool, S., & Koul, R. (2011). Self-determination theory and career aspirations: A review of literature. *Institutions, 7*, 9.

Chaus, O. (2019). Artificial Intelligence and robotics. In *Modern technologies: Improving the Present and impacting the future.* Дніпровський національний університет залізничного транспорту імені . . .

Chung, C. J. C., Cartwright, C., & Cole, M. (2014). Assessing the impact of an autonomous robotics competition for STEM education. *Journal of STEM Education: Innovations and Research, 15*(2).

Dictionary. (n.d.).

Feigenbaum, E. A., & Feldman, J. (1963). *Computers and thought.* New York McGraw-Hill.

Frey, C. B., & Osborne, M. A. (2013). *The future of employment: How susceptible are jobs to computerisation?.* Amsterdam: Elsevier.

Grant, A. M., & Parker, S. K. (2009). 7 redesigning work design theories: The rise of relational and proactive perspectives. *The Academy of Management Annals, 3*(1), 317–375.

Gray, D. A., Gault, F. M., Meyers, H. H., & Walther, J. E. (1990). Career planning. *Prevention in Human Services, 8*(1), 43–59.

Howaldt, J., Kopp, R., & Schultze, J. (2017). Why Industrie 4.0 needs workplace innovation—A critical essay about the German debate on advanced manufacturing. In *Workplace innovation* (pp. 45–60). New York: Springer.

Kang, H. S., Lee, J. Y., Choi, S., Kim, H., Park, J. H., Son, J. Y., . . . Do Noh, S. (2016). Smart manufacturing: Past research, present findings, and future directions. *International Journal of Precision Engineering and Manufacturing-Green Technology, 3*(1), 111–128.

Keqiang, L. (2016). *Full text: Report on the work of the government (2015).* Government of Canada.

Kiel, D., Müller, J. M., Arnold, C., & Voigt, K.-I. (2017). Sustainable industrial value creation: Benefits and challenges of industry 4.0. *International Journal of Innovation Management, 21*(8), 1740015.

Kooij, D. T. A. M., & Kanfer, R. (2019). Lifespan perspectives on work motivation. In *Work across the lifespan* (pp. 475–493). Amsterdam: Elsevier.

Lent, R. W., & Brown, S. D. (2008). Social cognitive career theory and subjective well-being in the context of work. *Journal of Career Assessment, 16*(1), 6–21.

Lent, R. W., Lopez Jr, A. M., Lopez, F. G., & Sheu, H.-B. (2008). Social cognitive career theory and the prediction of interests and choice goals in the computing disciplines. *Journal of Vocational Behavior, 73*(1), 52–62.

Matson, E., DeLoach, S., & Pauly, R. (2004). Building interest in math and science for rural and underserved elementary school children using robots. *Journal of STEM Education: Innovations and Research, 5*(3).

McCarthy, J., Minsky, M. L., Rochester, N., & Shannon, C. E. (2006). A proposal for the dartmouth summer research project on Artificial Intelligence, August 31, 1955. *AI Magazine, 27*(4), 12.

Müller, J., Maier, L., Veile, J., & Voigt, K.-I. (2017). Cooperation strategies among SMEs for implementing industry 4.0. In *Digitalization in supply chain management and logistics: Smart and digital solutions for an industry 4.0 environment. Proceedings of the hamburg international conference of logistics (HICL)* (Vol. 23, pp. 301–318). Berlin: Epubli GmbH.

Schaub, M., & Tokar, D. M. (2005). The role of personality and learning experiences in social cognitive career theory. *Journal of Vocational Behavior, 66*(2), 304–325.

Sevindir, H. K., & Yazici, C. (2014). Examining the factors affecting the selection of mathematics profession: A case study. *Procedia-Social and Behavioral Sciences, 152*, 642–647.

Shabbir, J., & Anwer, T. (2018). Artificial Intelligence and its role in near future. *ArXiv Preprint ArXiv:1804.01396.*

Shahzad, K. (2018). Boosting Pakistan with Artificial Intelligence. *Daily Times.*

Sheldon, K. M., Holliday, G., Titova, L., & Benson, C. (2020). Comparing holland and self-determination theory measures of career preference as predictors of career choice. *Journal of Career Assessment, 28*(1), 28–42.

Siau, K., & Yang, Y. (2017). Impact of Artificial Intelligence, robotics, and machine learning on sales and marketing. *Twelve Annual Midwest Association for Information Systems Conference (MWAIS 2017)*, 18–19.

Silk, E. M., Higashi, R., & Schunn, C. D. (2011). Resources for robot competition success: Assessing math use in grade-school-level engineering design. In *American society for engineering education*. Washington, DC: American Society for Engineering Education.

Teng, X. (2019). Discussion about Artificial Intelligence's advantages and disadvantages compete with natural intelligence. *Journal of Physics: Conference Series, 1187*(3), 32083. IOP Publishing.

Wisskirchen, G. (2017). *Artificial Intelligence and robotics and their impact on the workplace.* IBA Global Employment Institute.

12 Transitional Changes towards Flipped Classroom Approaches amidst the COVID-19 Pandemic to Develop Online Learning Communities

Muhammad Mujtaba Asad, Roha Athar, Irfan Ahmed Rind, Imran Khan and Al-Karim Datoo

CONTENTS

12.1 INTRODUCTION

This research study examines a paradigm shift of significant transformation in education and technology. Great progress can be observed in the educational field. Revolutionary advancements have cause teachers, students and parents to use technologically-based approaches at home as well as at school. "Flipped classroom" is one of the innovative instructional approaches re-shaping classroom teaching and learning. It consists of two segments: in the first, students watch lessons and videos at home, and in the second phase, the teacher organizes an interactive session inside the classroom. This concept has been evolving in the educational field over the past five years. Hence, the standard classroom pedagogy is flipped. In the traditional model, students prepare for the next lecture by reading textbooks at home. However, the flipped classroom model involves technology-based approaches which include instructional audio and video resources for real-world problem solving. Hence, flipped classroom opens new doors for student's learning. Now the question arises, "Is the teacher's role is diminishing, and are students more overloaded?" The only answer to this question is that the role of the teacher switched from lecturer to facilitator, and students are becoming more engaged and active recipients of knowledge.

The current outbreak of coronavirus known as Covid-19 was first identified in Wuhan, China and the WHO (World Health Organization) soon declared it a public health emergency to combat the spread of this disease. UNESCO is playing its vital role in facilitating countries in continuing the education through remote learning (UNESCO, 2020b). In this situation, must engage their administration, staff, faculty, and students in online teaching and learning without compromising their functioning and while keeping their staff and learners safe from this unseen virus. In this pandemic situation, a type of blended learning we call "flipped classroom" is one of the strategies used by teachers (Hodges et al., 2020).

In a flipped classroom, students prepare themselves before class by watching instructional videos sent by teachers, and during classtime, the content is interactively discussed. In this way, the teacher is better able to engage students in hands on activities, problem-solving and so on. During Covid-19, the flipped classroom as an instructional strategy is being used by most teachers. Online teaching resources are becoming widely used, promoting active learning, collaborative learning, engagement, motivation, peer-assisted learning and self -directed learning, as well as improving ICT skills (Akçayır & Akçayır, 2018). This study was conducted to recognize gaps in the literature and to identify good practices through the use of flipped classroom learning amidst the Covid-19, as well as to identify challenges faced by teachers in using flipped classroom.

12.2   COVID-19 TRANSMISSION

The highly contagious coronavirus that causes the disease Covid-19 has taken many lives. Its symptoms include fever, flu, dizziness, respiratory infection, and in some cases, pneumonia (Kenneth McIntosh, 2020). The WHO stated that it spreads when an infected person sneezes or coughs, thus transmitting droplets. Due to this pandemic situation, the WHO declared it a global public health emergency. Lockdowns have meant closure of schools, banks, markets, industries, and other workplaces, which have negatively impacted employment, physical learning and teaching, human psychology, and national economies (Tahir & Masood, 2020).

12.3   IMPACT OF COVID-19 ON EDUCATION

As discussed by many researchers, Covid-19 is a contagious disease spreading through human contact, an invisible and mighty enemy passing easily from one person to another (UNESCO, 2020). To limit cases, governments have made decisions to shut down. A 2020 article in *Education International* declared that due to Covid-19, the global education system has been hit very hard, affecting 1.5 billion children and 63 million teachers (Education International, 2020).

A US report recommended school closures of in-person teaching and learning. This report further discussed the role of UNESCO in this Covid-19 pandemic situation, which caused disruption in K-12 as well as higher education (Educational week, 2020). UNESCO has set an exemplary opportunity for teachers, students, parents, and other stakeholders through the commencement of distance learning (UNESCO, 2020a). Pradeep Sahu (2020) advocated that this epidemic situation has forced teachers to shift from face-to-face teaching to online teaching. However, the use of computers and laptops are becoming more necessary nowadays for parents, teachers and students to stay connected. Thus, it is going to be difficult for some faculty and parents who are not tech-savvy (Sahu, 2020).

The UNICEF Global Chief of Education proposed some tips for parents to help their children to learn at home so that they participate more actively in online classes. It suggests parents make a routine to engage children in activities that serve as learning opportunities. It advised parents to engage children in open conversation and provide opportunities to express their feelings easily. It also recommended that parents must be aware of activities students are doing through digital platforms and also to stay in touch with teachers (UNESCO, 2020c). Further researchers supported this concept and stated that parents or guardians should supervise the activities of young learners. Moreover, teachers should be in contact with parents to update them about activities and student performance (Ibo, 2020). Now the role of teacher runs parallel to that of a parent's. A research study recommended some guidelines for teachers working in schools. The International Baccalaureate Organization (2020) identified some techniques to work with more authenticity. Use of formative assessment, real-life applications, inquiry-based learning and collaborative, cooperative group discussions are recommended in this article (Ibo, 2020). The review of research studies showed that disadvantaged families are suffering more in this pandemic; this transition is creating inequalities and increasing society's digital divide. It claimed that most countries are stiff suffering and have not started online teaching due to lack of resources. The importance of physical classrooms cannot be replaced by alternative modes of learning (Education International, 2020).

12.4 EMERGENCY REMOTE TEACHING

In comparison to online teaching, emergency remote teaching is a temporary transformation of instruction delivery to a substitute mode, specifically in crisis. The main aim is to re-organize educational setups to provide feasible remote teaching solutions in an emergency. Under the heading of Emergency Remote Teaching, many examples of schools and universities implementing technology-based approaches like mobile learning, blended learning, flipped classroom and so on are provided by Hodges et al. (2020). In the current Covid-19 situation, a study reported that school closures are a turning point for the education sector of all affected countries. Most countries have released comprehensive details and recommendations about options regarding digital resources, grading policies, and scheduling for conducting classes (Slama, 2020).

Recent research on the concept of virtual schooling and home schooling has been given by Fitzpatrick and Ohio. Additionally, Molnar (2019) discussed how teachers spends five to six hours per week to give instructions. They observed that the teachers with the best digital skills spend most of their time reaching out to students and their families, providing instructions, coaching and making extra efforts keep all students learning in pace (Molnar et al., 2019). Other studies focused on the major shift in this pandemic situation that had important implications for remote learning, and others highlighted the important role of schools in identifying needs and challenges faculty and students face (Slama, 2020). This shift is also disrupting the ability of students and staff to maintain and update instructions on time. Asynchronous activities are suggested as reasonable. It is to keep in mind that students are not supposed to respond to the lecture immediately, but they have to wait for during class activities in flipped classroom (Hodges et al., 2020). In summary, the flipped classroom model is an emerging strategy to fight against Covid-19 while continuing with education. It is best to provide instructions to students prior to online sessions to allow for more comprehensive discussion.

12.5   FLIPPED CLASSROOM APPROACH

The flipped classroom approach is an emerging concept in the field of education, gaining popularity day by day in promoting a student-centred approach. Researchers have defined the flipped classroom as an "instructional strategy in which out-class timing is used for pre-recorded video lectures and in-class timing is used for collaborative activities based on problem solving, discussions, critical thinking and other group works" (DeLozier & Rhodes, 2017). The concept of the flipped classroom was put forth by Jonathan Bergmann and Aaron Sams, who initially set to help students who missed in-class activities for any reason. They made a series of online video lectures to watch while at home. Sams asked students to watch it, and then they used in-class time for discussions and collaborative group activities. This approach of teaching and learning flipped the concept of traditional classrooms (Drake et al., 2016). According to Bergmann and Sams (2012), the most essential tool for flipped classroom is technology. For the access of online pre-recorded videos, students need technology at home. In the same manner, teachers also require technological tools to record, edit and publish video lectures. It demands planning, time, commitment and effort for successful online content delivery. Furthermore, the length and information shared in video must be appropriate to the students' capacities, age levels, and attention spans, so they may engage actively. The good quality of video content leads to a successful flipped classroom (Bergmann & Sams, 2012). In today's Covid-19 pandemic situation, many teachers have chosen video lectures as opposed to in-person lectures due to closures of traditional physical classes. A study conducted by Lipomi (2020) showed the difficulty teachers are struggling with along with explaining the importance of online resources in Covid-19. Lipomi (2020) showed that teachers favoured video for promoting active remote learning. This approach is convenient for students because they can skip forward, pause, and rewind the uploaded lectures (Lipomi, 2020).

Much research has advocated the superiority of flipped classrooms over traditional instructional strategies. This approach has altered the traditional pedagogical order of teaching and learning by emphasizing learning more than teaching and incorporating hands-on practical activities during class (Wang, 2017). The combination of digital options and a flipped classroom approach provides a platform for enhancing both learning and pedagogical skills. Integration of technology provides opportunities for students to access material even outside school facilities. This transformation in teaching and learning has shown positive impact in the educational world. Video is a significant and effective multimedia tool that merges images, text and sound. Research has also demonstrated that students find video a more effective and comfortable way to understand and use their critical thinking skills as opposed to traditional methods of reading textbooks (Lipomi, 2020).

Literature showed that flipped classroom has opened doors for more enriched learning activities like hands-on and problem-solving activities. It makes students responsible for their own learning. Millard discovered five main ways through which flipped classroom works. It increases student engagement, strengthens team-based skills, offers guidance, promotes classroom discussion, and offers opportunities for teachers. Furthermore, a study reported that 99% of teachers are using flipped classroom in junior and secondary schools. Empirical results showed the following percentages for how often certain subjects are being taught via flipped classroom; science at 46%, math at 32%, and English language at 12%. In addition, 67% of students showed improvement in their learning and 80% students boosted the engagement. Online resources utilized by math teachers include Khan Academy, through which students watch videos, practice sums, and take quizzes. Similarly, science teachers use online as well as offline tools. They engage students in watching videos on YouTube or email the material to them. They enrich students' understanding by employing them in activities like open-ended questions, graphic organizers, and sentence framing (Schmidt & Ralph, 2016).

A review of the studies advocates the extensive use of technology in the field of education of developing countries. Yang (2014) conducted a research in Chinese elementary and secondary schools favoured the flipped classroom as it brought revolutionary changes as opposed to traditional classroom changes. It is revealed that Chinese students familiar with tradition classroom learning were not competent at cooperative activities as substitutes to individual learning. This is because previously, teachers worked as knowledge transmitters, and students were only there to absorb that knowledge. But this approach of the flipped classroom has changed everything, including teaching methods, learning activities, evaluation, and assessment (Yang, 2014).

Cooperative learning is one of the methods in which learners work for one goal and widen one's individual and group competencies. It promotes interaction and emphasizes individual as well as group accountability. It requires small group intrapersonal skills that function to improve group effectiveness. The flipped structure in Chinese classrooms has allowed students to watch videos before class and then during class, engage students in group learning activities which facilitate teamwork and problem-solving skills. Assigning different tasks to each group member makes students responsible for their own learning. This model proves to be good for Chinese students' motivation (Yang, 2014).

Reviews of studies about the use of the flipped classroom approach unveiled that teachers plays a crucial role in learning activity. They also highlighted the importance of parents' and peers' roles in creating meaningful learning environments. As the flipped classroom advocates group learning activities, teachers must guide and facilitate group leaders to manage and organize activities by enhancing 21st-century skills (Wasriep, 2019). Another research article explored the viewpoints of Asian universities as regarding flipped classroom. Culture is a dynamic element which shape an individual's beliefs. This article highlighted the contrast between Western and Asian teachers' beliefs which directly impact their teaching philosophies. Many Asians consider students passive learners, while Western teachers hold the opposite view. Asian students view their teachers as sources of information on which they rely highly for successful test scores. On the contrary, Western students prefer to learn independently. The inverted classroom approach places the teacher's role as a facilitator by the side of students. This shift is causing students anxiety, and teachers are not able to handle it due to adhered traditional beliefs. The result analyzed from the nine Asian countries studied suggested continuing this unconventional model, as it is generating an effective learning environment, enhancing students learning and emphasizing active learning. Furthermore, many students understand the concept of globalization in this modern technological world where information is at their fingertips, and they consider teachers as secondary sources. While supporting flipped classrooms, students additionally focused on soft skills to be successful in the future (Chua & Lateef, 2014). Another study proposed flipped classroom as a solution to traditional hierarchical beliefs of teachers and students. The conclusion reported that flipped classroom positively affects students' motivation and engagement in English writing tasks.

Moreover, interactive teaching environments and satisfaction from the students' side has proven flipped classroom as an innovative approach, which also allows teachers to spend more time helping solve students' individual learning problems. Literature suggests that a lack of reasoning and critical thinking skills prevents students from becoming proficient and competent. One reason which causes deficiency in these skills is solely attending boring lectures in face-to-face classes. Flipped classroom is a solution to this issue because it focuses on constructivism, which says "individuals use their prior knowledge to experience new concept. In a like manner, author also valued the importance of flipped classroom as it plays vital role in enhancing higher order cognitive skills" (Sultan, 2018).

12.6 IMPORTANCE OF FLIPPED CLASSROOM IN ENHANCING STUDENTS' ENGAGEMENT/ LEARNING

The flipped classroom approach is gaining in popularity and importance in the educational process. Day (2018) conducted an experimental study in Boston comparing two groups of students: one was traditional and the other was experimental, using a flipped classroom approach. Significant results showed that group used flipped classroom performed better in their final grades in contrast to students in traditional ones (Day, 2018). A review of studies revealed many benefits of the flipped classroom approach. According to Awidi and Paynter (2019), flipped classroom as a pedagogical approach proves to be a transformational model in enhancing student engagement, improving the learning experience, and eventually improving student outcomes.

The use of the flipped classroom approach supports meaningful construction of active knowledge and experiential learning. Some researchers have also identified improvements in social learning in the practical use of flipped classroom. Further research has suggested that flipped classroom also proves to be a better option in enhancing student motivation, their ability to manage cognitive loads, and improve learning outcome (Awidi & Paynter, 2019). For instance, a study reported that more than half of the studies consider flipped classroom as one of the key components of imparting quality education, and it plays a significant role in improving learning outcomes of students measured by standardized test scores, GPAs, and other assessment strategies. It also enhances student motivation and level of engagement, which are important elements of a conducive learning environment. According to one of the reviewed studies, flipped classroom enhanced student's motivation up to 18%, and active engagement till 14%. In addition, it also highlighted some other areas which are improving in flipped classrooms, including satisfaction, confidence, active and meaningful knowledge, creativity, problem-solving skills, retention, application skills, and ICT skills. The same study focused on pedagogical contributions by the teachers in flipped classroom through the involvement of students in collaboration, personalization, higher order thinking, and self-directed learning. In this way, it provides flexibility to teachers as well as students, not only in terms of access but also enjoyment and satisfaction. The same study reported that groups of students taught through flipped classroom approaches have a 1% improvement in attendance and 1% reduction in course withdrawal. In the same manner, university students showed more positive attitudes towards the use of flipped classrooms as opposed to conventional methods, with 14% increment in the percentage for the student's perceptions at university level. Also, student-teacher and student-student interactions rose nearly 20%, because this model provides more time for such interactions. Finally, another advantage reported in this study is that flipped clasrooms contribute to a more efficient use of class time more efficiently, as lectures are read and watched outside of class, allowing more in-class time for student-centred activities like discussion, debate, feedback, hands-on activities, and so on (Akçayır & Akçayır, 2018). Another study evaluated the efficacy of flipped classrooms by conducting online quizzes, wrap-sessions, and discussion forums, and showed that this approach proved to be helpful in improving students' performance and attendance (Stöhr et al., 2020).

Shih and Tsai (2017) discovered another benefit of using flipped classroom in a marketing course by engaging students in online, project-based learning. Results showed that the model improved learning interest and effectiveness as well as teamwork (Shih & Tsai, 2017). Schmidt and Ralph (2016) conducted research on teachers who used flipped classroom in junior, high and secondary school. The study indicated 99% of teachers showed willingness to continue this approach, student performance improved 67%, and student engagement increased 80% (Schmidt & Ralph, 2016). Another study suggested that flipped classroom makes learners responsible for their own learning. Students feel free to use other online resources for more comprehension and enjoy online lectures because they have sufficient time. A new aspect found in this study is that this approach also engages parents in contributing towards their childrens' learning. The study showed that parental involvement in students' homework improves students' academic achievement (Chua & Lateef, 2014).

Cognitive load theory states that it is difficult for working memory to retain information. In this regard, another benefit of the flipped classroom model is that it provides students an opportunity to review content again whenever they need and take notes of important information. It is also mentioned in this report that the model increases student collaboration, presentation skills, problem-solving skills and development of other skills. It also has the unique advantage in that it also educates parents, who can watch video lectures with their child (Gilboy et al., 2015). Further, this study claimed that the flipped model not only engages students in class time but also outside of class. Quick supply of formative feedback as well as improved concentration on learning are further positive characteristics of the flipped classroom approach (Asif & Omer, 2018).

12.7 CHALLENGES IN USING FLIPPED CLASSROOM

Undoubtably, the flipped classroom approach has changed traditional educational models, teaching methodologies, students' learning styles and so on. On one hand, it offers a range of flexible benefits which are affecting different aspects of teaching and learning in a positive way. On the other hand, this model has also presented some challenges, reported in one research article that claimed that this model is massively dependent on technological tools and internet access, making it difficult for teachers who do not have internet access to upload video lectures, and also problematic for students who do not possess electronic devices at their home (Gilboy et al., 2015). Following this limitation, another study supported that due to the absence of technological tools, many students are not able to prepare before class and thus engage actively in class activities. Technological competency is also considered troublesome for teachers; video lectures with poor quality, limited pedagogical skills, and technical issues affect students' learning negatively. In addition, it is also reported that there is an inverse relation between length of the video and videos viewed by students. Obviously, insufficient technological competency causes poor and low student learning outcome, and ultimately impacts the efficacy of flipped classrooms (Akçayır & Akçayır, 2018; Chua & Lateef, 2014). It is also difficult for novice teachers to actively engage students during video watching. They need extra efforts and time to demonstrate how to take notes of important information during watching videos (Gilboy et al., 2015). While exploring students' perspectives about flipped classroom, it is revealed that students feel an extra burden, as this model requires that students watch or read lectures outside of class time, which many students do not find feasible, and thus do not prefer this model, preferring traditional classrooms. These students experience anxiety, resistance to change and adoption problems with this model (Akçayır & Akçayır, 2018). Another study shared some limitations of the flipped classroom approach, including teachers feeling burdened with utilizing a lot of material, which can feel boring. It was also frequently mentioned that students do not consistently complete their homework, resulting in low academic outcomes. Moreover, research conducted in China also highlighted some existing problems in carrying out flipped classroom. It claimed that flipped classroom learning leads to poor performance due to poor video and audio graphics. In this manner, teachers who are not competent enough to use graphics, edit, and compose them, caused a decrease in student interest and low learning efficiency. Similarly, a poor classroom structure and design is responsible for creating problems for teachers. In traditional classrooms, they engage students through lectures and other strategies easily, but it is difficult for teachers to engage students in other activities besides answering their queries from videos watched already at home.

A growing number of studies also have also pointed to the active participation of parents in monitoring their child's activities, which leads to higher student grades. Correspondingly, parents who are not well-educated and are not aware of this new interactive model are not able to provide the same support. With less paperwork, it can be more difficult for parents to assess student performance (Akçayır & Akçayır, 2018). Equally important to mention is a study indicating negative perceptions of students regarding the flipped classroom approach; they used the phrase "Teaching ourselves" to describe a lack of guidance and instruction. Making students responsible for their own learning has caused fear among students due to uncertainty about their success. They reported excessive workloads and claimed that this model may be feasible and suitable for teachers, but does not impact all students in a positive way. However, students of disciplines like business, medicine and nursing showed willingness to continue this approach. One of the difficulties for teachers is in choosing appropriate strategies for class time from a number of activities, and redesign them according to student interest, age, and level. Another very interesting issue which is pointed out by a growing number of research studies is how this model evaluates and assesses problem-solving and critical thinking skills. As literature indicates, this paradigm shift demands a change in traditional beliefs strongly held by teachers and students (Rotellar & Cain, 2016). An analysis of reviewed studies identifies three main themes: students' perceptions, facultys' perception and other operational challenges. Data related to students' perception showed that students need more clear guidance to work in groups. When students felt doubts during video watching, they were not able to ask questions during the videos. Teachers using limited resources did not value the flipped classroom approach. Operational challenges included no internet access, difficulty in ensuring that students had watched videos, and lack of ICT skills (Lo & Hew, 2017).

12.8 CONCLUSION

Once Benjamin Franklin stated, "Tell me and I forget, teach me and I may remember, involve me and I learn". This is the core concept behind the flipped classroom approach. It provides a range of benefits investing in teachers' pedagogical growth and empowers students to develop higher cognitive skills. As noted earlier, the flipped classroom model enhances students' engagement, improves their attendance, learning experiences, outcomes, as well as supports meaningful construction of active knowledge and experiential learning, satisfaction, confidence, creativity, problem-solving skills, retention, application skills, and ICT skills. On the contrary, there are limitations to implementing a flipped classroom approach in developing countries, such as: less competency of IT skills, change in mindsets, lack of resources and knowledge of new strategies, limited pedagogical skills, technical issues that affect learning negatively, extra burden on students, anxiety, resistance to change and adoption problems, poor classroom structure, parents who are not well-educated and are not aware of this new interactive model, fear among students due to uncertainty about their success and workload, optimization of scheduling class times, difficulty in evaluating and assessing problem-solving and critical thinking skills, no internet access, difficulty in ensuring student participation, lack of ICT skills, and so on. The importance of physical classrooms cannot be replaced by alternative modes of learning. However, the use of computers and laptops are becoming more necessary nowadays for parents, teachers, and students to be stay connected. Despite these drawbacks, this model is commonly used by educators amidst the Covid-19 pandemic situation. To overcome this crisis, teachers are shifting towards online teaching, uploading lectures on YouTube, sending PowerPoint presentations, and other strategies. The review of research studies showed that disadvantaged families are suffering more in this pandemic; this transition is creating inequalities and increasing the digital divide in the society. Rapid advancement in technology caused flipped classroom to be implemented globally. The implementation and execution of this new strategy demands transformations in the traditional beliefs and teaching philosophies of teachers and students, as well as vast changes in educational environments, class sizes, human and technical resources, important aspects of curriculum, appropriate instructional strategies, and more.

REFERENCES

Akçayır, G., & Akçayır, M. (2018). The flipped classroom: A review of its advantages and challenges. *Computers & Education, 126,* 334–345.

Awidi, I. T., & Paynter, M. (2019). The impact of a flipped classroom approach on student learning experience. *Computers & Education, 128,* 269–283.

Bergmann, J., & Sams, A. (2012). Before you flip, consider this. *Phi Delta Kappan, 94*(2), 25–25.

Chua, J. S. M., & Lateef, F. A. (2014). The flipped classroom: Viewpoints in Asian universities. *Education in Medicine Journal, 6*(4).

Day, L. J. (2018). A gross anatomy flipped classroom effects performance, retention, and higher-level thinking in lower performing students. *Anatomical Sciences Education, 11*(6), 565–574.

DeLozier, S. J., & Rhodes, M. G. (2017). Flipped classrooms: A review of key ideas and recommendations for practice. *Educational Psychology Review, 29*(1), 141–151.

Drake, L., Kayser, M., & Jacobowitz, R. (2016). *The flipped classroom. An approach to teaching and learning.* Available at: www.newpaltz.edu/media/the-benjamin-center/P.Brief_2020Vision-Flipped%20classroom.pdf (Accessed: 15 May 2020).

Education International. (2020). *COVID-19: Educators call for global solidarity and a human-centred approach to the crisis.* Available at: www.ei-ie.org/en/detail/16723/covid-19-educators-call-for-global-solidarity-and-a-human-centred-approach-to-the-crisis (Accessed: 25 April 2020).

Education Week. (2020). *Map: Coronavirus and school closures.* Available at: www.edweek.org/ew/section/multimedia/map-coronavirus-and-school-closures.html (Accessed: 15 May 2020).

Gilboy, M. B., Heinerichs, S., & Pazzaglia, G. (2015). Enhancing student engagement using the flipped classroom. *Journal of Nutrition Education and Behavior, 47*(1), 109–114.

Hodges, C., Moore, S., Lockee, B., Trust, T., & Bond, A. (2020). The difference between emergency remote teaching and online learning. *Educause Review, 27.*

International Baccalaureate Organization. (2020). *Online learning, teaching and education continuity planning for schools.* Available at: www.ibo.org/globalassets/news-assets/ coronavirus/online-learning-continuity-planning-en.pdf (Accessed: 27 April 2020).

Lipomi, D. J. (2020). Video for active and remote learning. *Trends in Chemistry.*

Lo, C. K., & Hew, K. F. (2017). A critical review of flipped classroom challenges in K-12 education: Possible solutions and recommendations for future research. *Research and Practice in Technology Enhanced Learning, 12*(1), 4.

McIntosh, K. (2020). *Coronavirus disease 2019 (COVID-19): Epidemiology, virology, clinical features, diagnosis, and prevention.* Available at: www.uptodate.com/contents/coronavirus-disease-2019-covid-19-epidemiology-virology-clinical-features-diagnosis-and-prevention (Accessed: 24 April 2020).

Molnar, A., Miron, G., Elgeberi, N., Barbour, M. K., Huerta, L., Shafer, S. R., & Rice, J. K. (2019). *Virtual schools in the US 2019.* National Education Policy Center.

Rotellar, C., & Cain, J. (2016). Research, perspectives, and recommendations on implementing the flipped classroom. *American Journal of Pharmaceutical Education, 80*(2).

Sahu, P. (2020). Closure of universities due to coronavirus disease 2019 (COVID-19): Impact on education and mental health of students and academic staff. *Cureus, 12*(4).

Schmidt, S. M., & Ralph, D. L. (2016). The flipped classroom: A twist on teaching. *Contemporary Issues in Education Research (CIER), 9*(1), 1–6.

Shih, W. L., & Tsai, C. Y. (2017). Students' perception of a flipped classroom approach to facilitating online project-based learning in marketing research courses. *Australasian Journal of Educational Technology, 33*(5).

SLAMA, J. R. J. B. F. H. H. L. L.-T. M. N. T. (2020). *Remote learning guidance from state education agencies during the COVID-19 pandemic: A first looks.* Available at: https:// edarxiv.org/437e2 (Accessed: 5 May 2020).

Stöhr, C., Demazière, C., & Adawi, T. (2020). The polarizing effect of the online flipped classroom. *Computers & Education, 147*, 103789.

Sultan, A. S. (2018). The flipped classroom: An active teaching and learning strategy for making the sessions more interactive and challenging. *Journal of Pakistan Medical Association, 68*(4), p. 630.

Tahir, M. B., & Masood, A. (2020). *The COVID-19 outbreak: Other parallel problems.* Available at SSRN 3572258.

UNESCO. (2020a). *COVID-19 educational disruption and response.* Available at: https:// en.unesco.org/covid19/educationresponse (Accessed: 24 April 2020).

UNESCO. (2020b). *Pakistan COVID-19 weekly situation report no. 2.* Available at: https://relief-web.int/sites/reliefweb.int/files/resources/UNICEF%20Pakistan%20CoViD-19%20 Situation%20Report%20No.%202%20-%2030%20March-5%20April%202020.pdf (Accessed: 25 April 2020).

UNESCO. (2020c). *5 ways to help keep children learning during the COVID-19 pandemic.* Available at: www.unicef.org/coronavirus/5-tips-help-keep-children-learning-during-covid-19-pandemic (Accessed: 25 April 2020).

Wang, T. (2017). Overcoming barriers to 'flip': Building teacher's capacity for the adoption of flipped classroom in Hong Kong secondary schools. *Research and Practice in Technology Enhanced Learning, 12*(1), 6.

Wasriep, M. F. (2019). Enhancing the 21st century learning through the flipped classroom approach: A science teacher's S perspectives. *Asia Proceedings of Social Sciences, 4*(2), 121–124.

Yang, J. (2014, May). Implementing the flipped classroom in elementary and secondary schools in China. In *International conference on education, language, art and intercultural communication (ICELAIC-14).* Atlantis Press.

Index